UNIVERSE ALTERNATIVES

EMERGING CONCEPTS OF SIZE, AGE, STRUCTURE AND BEHAVIOR

Billy L. Farmer

Publisher's Cataloging in Publication
(Prepared by Quality Books Inc.)

Farmer, Billy L.
Universe alternatives / emerging concepts of size, age, structure and behavior / Billy L. Farmer.
p. cm.
ISBN 0-9649983-3-5

1. Cosmology. 2. Particles (Nuclear physics) 3. Relativity (Physics) I. Title.

QB981.F37 1996 523.1
QBI96-20124

Library of Congress Catalog Card number: 96-96145

Printed in the United States of America

For information about the availability of this book, please contact the author — B. L. Farmer, 6044 Bel Mar, El Paso, Texas 79912.

CONTENTS

PREFACE

The following writing is a very ambitious attempt to rethink the structure and behavior of the universe from its gross features down to its microscopic underpinnings. This material should appeal to those inquisitive individuals who have learned to maintain a healthy level of skepticism concerning all theoretical matters. It is understood and accepted, however, that my effort to reach a broad spectrum of readers will probably encounter varying degrees of resistance among those who see this project as an invasion of hallowed turf by an undocumented boat rocker.

The book can roughly be divided into two categories — total universe or cosmological speculations and a more basic section concerning simplified redefinitions of such things as light, mass, space and natural forces. This preface will be used to outline and summarize these items in the following paragraphs.

The Cosmology phase of the writing describes the large scale theoretical speculations that should come to mind as the Big Bang — expanding universe concept is replaced by an overall static model that will most probably be envisioned as being infinite in both size and age.

As a preliminary to listing and causally justifying the other linked sequential arguments involving light, space, mass, etc., it seems appropriate to first describe an important baseline type of assumption concerning the make-up of the universe.

This central and essential premise holds that the volume of the universe is completely occupied by a single substance or entity having a very wide range of perceived forms, of which mass and space are the two named examples. Of course, the significant implication of this concept is that space is not "empty" — rather, that space and mass are only two different phases or perceptions of a single universal entity. Consequently, I would suggest that, rather than taking the idea of emptiness too literally, it might be wiser to say that the current technological extensions of our senses are not yet sophisticated enough to discern the presence of a substance or entity that may inhabit a volume containing no detectable mass. The repeated lessons of history would seem to support this type of approach. *(Recall that about ninety years ago the contemporary scientific opinion shapers threw out the baby with the bath water when they decided to replace the old "ether" space medium theory with the concept of emptiness, simply because the Michelson-Morley class of light velocity experiments had discredited the somewhat magical aspects of the ether's assumed behavior.)*

In the single entity universe, light through space must be seen as a spherical wave like disturbance in the medium or phase of the universal entity that we call space; thus the speed of light must necessarily be a media dependent value. This means that light speed determinations by different observers should vary according to each observer's velocity relative to his surrounding space medium.

It is further proposed that the system of separately named forces should be replaced by a single type of universal cohesion having a very wide range of strength values and operating throughout all phases of the single universal entity. So, in addition to the commonly recognized gravitational attraction between quantities of mass, there would also be extremely minute values of cohesion and compression between and among volumes of space as well as between mass and space. However, the apparent repulsion of two properly oriented magnetic objects seems to be at odds with this idea of universal cohesion. One possible resolution of this inconsistency is given on pages 100-102. Briefly, the idea is that directional grouping of the internal motion patterns (domains) in each of these two precisely opposed magnetic objects allows them to mutually shield each other from a small angular segment of the surrounding universal cohesion — thus they tend to move apart in response to differential cohesion rather than a pushing force.

In the chapter on sub atomic structure, the validity of the standard type of intra-atomic architecture and behavior is questioned regarding most of its aspects. The usual concept is one of "discrete" quantities of mass and charged mass maintaining orbits in a matrix of empty space by balancing centrifugal force against "electrical" attraction. The mathematical construction of this standard model depicts orbiting particles and charged entities as being spatially very minute compared to the volume of empty space within the atom.

I would suggest an alternative picture in which the entire intra-atomic volume is completely occupied by a homogenous pre-mass type of substance characterized by patterns of natural and perpetual curved motion. The repeating nature of these motion geometries causes a vague subdivision into multiple contiguous units, which are further arranged into groups resembling a containment type of hierarchy organization (systems within systems). The strength of cohesion between any two contiguous units, groups, or volumes would be a function of the motion parameters within each of those volumes and the average distance between those motion patterns.

This particle physics chapter also contains several specific propositions and models concerning intra-atomic behavior and structure. These hypothetical examples are admittedly inaccurate and far from reality. They are included only for the purpose of promoting directional thinking toward a realistic atomic model that is compatible with a single entity (no empty space) universe.

From the theoretical framework outlined in the preceding paragraphs, a sequence of logic emerges that automatically questions the reality status of such things as emptiness, light speed assumptions, the qualitative segregation of natural forces, and the accepted models (Bohr, Rutheford, Thomson) of the atomic interior. Of course, this also leads to skepticism about black holes and both special and general relativity.

On the cosmic scale, it is suggested that the universe is an unlimited expanse of hierarchy organization, or systems within systems, in which there is a perpetual equilibrium maintained by cyclic concentrations and expansions at multiple hierarchy levels. This is pictured as a logical alternative to the Big Bang — expanding universe model with its necessarily finite temporal and spatial aspects. In support of the above preference in universe models, there is some evidentiary discussion of red shifts, background radiation, Virgo's non-expansion and the infinity concept.

Of the several established theories that are questioned in this book, the Big Bang — expanding universe scenario is the one that is most likely to collapse within the near future. In addition to the tunnel vision view that doggedly insists on designating recession velocity as the only consideration in explaining Hubble's distance related inter-galactic red shifts, there is a relentless and continuing accumulation of complicated and highly improbably corollary events and structures that must pertain in order for the Big Bang — expanding universe model to maintain even a minimum of logical probability in the minds of cosmologists.

I am aware that the above list of proposed changes in theory and basic assumptions is very drastic and far reaching. If even some of these items come to be seriously considered in the next few years, the disruption in textbooks, careers, tenures and research programs would be severe. For these reasons, there is a great deal of reflexive and emotional resistance to ideas that are as revolutionary and disruptive as those contained in this writing.

The first question that comes to mind when a potential reader recognizes the magnitude of the anti-establishment heresy suggested in this book is: Who is the author? And what is the qualifying background of one who dares to question such a diversity of scientific assumptions that have become so hardened and institutionalized by many years of acceptance? There is no way to avoid the fateful interest sapping admission that the author is not even a professional member of the physics community, but rather, a sixty-six year old Dermatologist who, only in recent years, began to indulge a long delayed desire to delve into theoretical physics. This lack of academic standing is obviously a severe roadblock to serious consideration of the contained ideas by busy scientists. The necessary screening of the voluminous incoming information and misinformation is reason enough for many physicists to give a patronizing smile as they toss this book into the "crack pot" file. It is also understood and accepted that this writing will be viewed by some as a brash and unorthodox transgression by a poorly qualified outsider — a messenger that is both intrusive and improperly dressed. Or, as my sharp son-in-law chided in a classic one liner: "You're writing about things that you shouldn't even be thinking about." However, in spite of these appropriate criticisms, I still believe that this material contains reasonable arguments that should be seen by leaders and students in the field, in addition to including suggestions that should properly influence future investigative directions.

Among the several disciplines of physics, Cosmology and particle physics stand out as necessarily being much less objective than other fields. In particle physics, the elements of the model being considered are inherently too small to be observable, and conversely — too expansive in the case of Cosmology. Educated guessing seems to be the most appropriate designation for the basic assumptions that must be used to construct theoretical models in both of these relatively invisible specialties of physics. I am reminded of a parallel situation among the specialties of medicine. Dermatology, for example, has become much more scientific and objective in the last twenty years or so; however, it still places considerable diagnostic reliance on such things as visual microscopic appearance as well as the experience and clinical grouping skills of the old masters. This stands in contrast to the more complete objectivity seen in some of the newer specialties like Immunology, Endocrinology and Cardiology.

In what seems to be one of the frailties or faults of human thought evolution, we sometimes see shaky theories become transformed into unquestionable or highly probable realities that are "accepted by most leading scientists," based on nothing more substantial than years of repetition and media popularity.

Rather than seeking a profit, my primary objective and motivation in this literary project continues to be: reaching as many free thinking amateur and professional theorists as possible with these controversial ideas. In 1995, a number of books were sent to university physics departments, authors, libraries and science editors of various periodicals. Possibly my individual promotional efforts will eventually result in professional publication and some degree of bookstore distribution. This would further fulfill my main objective of reaching as many readers as possible by changing the project toward an economically self sustaining venture.

With an eye toward expanding my mailing list, the names and addresses of any scientists or non-professional enthusiasts who might be interested in this type of material would also be appreciated.

Comments from readers are welcome, especially from those who feel a spark of excitement when contemplating the size of the thinking revolution that is proposed in these pages.

B. L. Farmer
January 1996

INTRODUCTION

The following literary project is a collection or series of causally linked speculations and arguments about the nature of the universe. It is being written at this time because I anticipate the long overdue demise of the fatally flawed Big Bang — expanding universe concept of how everything began and developed. This approaching event, which could reasonably occur within a very few years, will open the field of Cosmology to a flood of new ideas.

The many problems, shaky assumptions, and bizarre rescue attempts that have been part of the Big Bang's bumpy history were discussed at some length in my 1993 manuscript entitled "Stuck in the Big Bang Quagmire, Sixty Years is Enough." It is obvious from the wording in that title that I am looking forward to the time when we can freely speculate on the nature of things without being theoretically hobbled by such an improbable monster as the Big Bang. The situation is somewhat analogous to waiting around for someone to mercifully pull the plug on the life support system so that some poor brain dead individual can make a graceful exit.

Before outlining my ideas for filling this impending theoretical void, let me apologize to the reader for a writing style that may at times seem difficult to read and assimilate. Since the manuscript was drafted, consolidated, edited, and rewritten in longhand, it tends to contain more than the average number of thoughts per line. Not to excuse my own responsibility in this matter, I would suggest, however, that the complicated and descriptively difficult nature of the topics contributes to the necessarily sluggish flow of information from the paper to one's gray matter. It is definitely not the type of material that would tend to build confidence in one's ability at speed reading or instant comprehension. Also, trying to discuss this array of linked and convoluted subjects in separate chapters requires considerable duplication in setting up each topic. So, the reader should understand that frequent repetitions may be more related to the interwoven nature of the material than to editorial laziness on the part of the author.

It may seem to some that I tend to write like a loose cannon, taking delight in attacking all the established and respected theories and theorists in recent history. Such is not the case, however. My wide ranging criticisms would be better described as following the bouncing ball of sequential logic, more in the spirit of letting the chips fall where they may as one thing leads to another, rather than a calculated or vindictive campaign to raze the icons of physics.

As the Big Bang fades in credibility, the scientific community will experience a somewhat disturbing theoretical vacuum. There will suddenly be no concrete answers to the questions of size, age, motion and structural origin of the universe. One would expect an emerging series of questions and implied solutions that might be characterized in the following manner.

With the exit of the expanding universe concept, we are left with one that is static and very much older than the previous age limit of 15 billion years. Such a static and ancient universe must necessarily be very large or infinitely large to preclude the gravitational collapse that would have befallen any very old finite sized universe.

As far as the duration of the universe is concerned, there is a constantly increasing number of structural and behavioral observations suggesting great age. Also, the concept of infinite size seems to be incompatible with a temporal beginning as long as one does not consider magical creation by an intelligent deity as a reasonable probability.

Again, presuming a very old universe, the observed fusion disappearance of mass into the space sink says that there must be continuous creation of mass from the space phase to maintain today's chemical equilibrium.

Using the same reasoning as above, it follows that even large structures like galaxies must necessarily disburse and reform in a perpetually cyclic manner to maintain permanent equilibrium at all levels.

Then, if the space entity or space sink is part of the mass cycle, its status of "emptiness" must be questioned, which suggests the concept of a single universal entity occupying the entire volume of the universe with no intervening empty space.

This, in turn, affects the theory of light transmission and the basic assumptions of special relativity.

A universe that has been grossly dynamic for an infinitely long time obviously has no intention of ever burning out or coming to a cool, dark, and placid end. This suggests that the natural base line status of the universe must be one of eternal motion at the smallest units of organization. In other words, there will never be less universe activity than our present base line equilibrium of energy exchange.

Considering the parameters of this basic small unit curved motion as a cause for gravity and universal cohesion leads to reconsideration of all forces and related theories like general relativity.

Thus, the preceding series of linked subjects and arguments is a fairly good outline of the curiosity and skeptical thinking that motivated the topics and most of the chapter headings of this current writing.

It should be clear from the preceding outline that my approach to this writing was one of following a sequence of deterministic arguments rather than a blanket condemnation of all established theories. Even so, the reader will surely recognize the enormity of the heresy that I have laid down in these pages. Since I have found reason to doubt the basic assumptions concerning the architecture and behavior of mass, space, and energy, the result is a natural chain of skepticism that focuses on the soundness of the models developed by almost all of the theoretical and mathematical giants from Newton to Stephen Hawking. As it happens, practically all of the modern theory builders have dipped into this contaminated well of assumptions to some degree.

I hope that this material will be read by a large number of people, but especially I wish for it to be considered by some of the experts in the field. This is no easy task since the professional publications and periodicals understandably and correctly maintain certain standards of material acceptability. Authors must have some minimum of physics education and academic standing. Also, the material must survive a referee's scrutiny concerning some degree of conformity with accepted theory, among other things. Obviously, I am destined to strike out on all three of the above acceptability tests. So, this manuscript is being printed and distributed at my own expense. I have no regrets or resentment at having to spread my opinions in this manner, although it seems that there should be some more accessible publication opportunity for enthusiastic amateurs. This is especially true for those who cannot afford the time or expense of the self publishing route that I have taken.

Theories seem to enjoy considerable built in protection

against being dumped or even criticized once they have achieved that haven of safety that accompanies acceptance by the scientific community. This situation exists, to some degree, because our particular system of restriction to publication makes it difficult for critics, or boat rockers, at all levels of expertise to reach influential professionals.

There seems to be a widespread attitude of mind numbing complacency that has accompanied sixty years of more or less general scientific acceptance of the Big Bang — expanding universe concept. Because of this popular mind set and for the purpose of justifying the ambitious universe speculations in the following chapters, it seems appropriate to borrow from my 1993 writing in order to review and summarize the growing number of observations and logical arguments that should hasten the inevitable demise of the Big Bang theory.

This unfortunate blind alley detour of sixty years duration had its beginning in 1929 when Hubble revealed his observations that there was a direct and linear relationship between the distance to other galaxies and the amount of red shift of the light coming from them. This applied more clearly to the medium and far distant galaxies rather than the close ones. Then, as a fateful first step in the wrong direction, it was generally assumed that recession velocity was the only mechanism that should be considered in explaining Hubble's observations. This set the stage for a series of more or less automatic theoretical deductions as follows. Recession of all galaxies from each other could only be explained by a universe that was expanding at a steady rate. Using Hubble's red shift values and distance estimates plus this expanding universe assumption meant that the present universe, as a sphere of something less than 30 billion light years in diameter, must have begun as an explosion at a single point about 15 billion years ago.

The following points of criticism of the expanding universe — Big Bang scenario will only be mentioned briefly at this time since each of them is more fully explained in either the current writing or in my 1993 monograph.

The uniform smooth background microwave radiation we receive equally from all directions has features that suggest a universe that is both much larger and much older than the range of possibilities under the Big Bang theory.

Several aspects of today's universe suggest a repeating cycle of mass action equilibrium that has been going on for a very long time at many size levels of organization. This would push the age of the universe far beyond the Big Bang age limit. For example, even one cycle of galaxy formation and dissolution should logically have a duration of hundreds of billions of years.

Red shifts and other observations have shown that the Virgo system (of which our galaxy is a suburban part) is actually a gravitationally contracting (rather than expanding) cluster of over 1000 galaxies spanning more than 60 million light years. Obviously, this is evidence that all galaxy clusters are probably gravitationally closing on themselves to some degree. Still, the Big Bang enthusiasts stick to the very unlikely claim that the behavior of this 60 million light year expanse is just an unrepresentative local aberration in an overall expansion.

There is a rapidly increasing number of observations indicating that the age of various stellar and galactic items is greater than the theoretical age of the universe. Accumulations of such unacceptable contradictions are going to be an important part of the avalanche of impossible situations that will eventually bury the theory.

Explaining the beginning as a point explosion 15 billion years ago has resulted in a really bizarre collection of contrived forces, particles, entity unit building schemes, and organizational behavior. The imaginative mind set needed to preserve this improbable theory is exemplified by the fairly recent ad hoc fabrication of a modifying interval in the early seconds called the "inflationary period" during which the proper turbulences supposedly occurred in order that galaxy formation could be conceivable.

Another aspect of this sequence that is usually ignored, dismissed as an improper discussion subject, or simply called a "singularity" that requires no explanation, is the question of how the mass of the universe arranged to have itself condensed into this primordial egg prior to the explosion. Would not compression rebound phenomena like super novae or galaxy expansions be triggered long before this degree of ultimate density was reached?

In a final criticism, recall that the entire Big Bang sequence of events was made necessary by the initial presumption that recession velocity was the only possible cause for Hubble's red shifts. Halton Arp and others have developed a body of evidence indicating that quasars and active galaxies have the ability to produce red shifts that have nothing to do with distance or recession velocity. (Because of this, all the standard values quoted for size, distance, universe distribution, and associative relations of quasars and active galaxies are in error and must eventually be revised. This is true because these values are based on the erroneous assumption that all red shifts are completely distance and recession velocity related.) At any rate, since the core assumption that red shifts must necessarily represent recession velocity is in error, then the necessity for a Big Bang should be questioned.

Thus the Big Bang is unlikely for a number of reasons, not the least of which is the extremely complicated and intricate nature of the required events that are demanded by an expanding universe. Of course, natural evolution or development can be extremely complicated in a setting of long term natural selection, but having such a myriad of necessarily distinctive forces, particles, pre-particles, entities and coordinated movements involved in a sequentially dependent series of events that must all cooperate in a one time highly accurate and uniform explosion to produce today's galaxy picture stretches imaginative credibility a bit too far.

Unfortunately, even those in the cosmology establishment who recognize the flaws will find it difficult to back out as this sand castle begins to crumble around them — too much fame, tenure, endowment, reputation and ego investment ride on the validity of the Big Bang theory. Thousands of pages of textbooks and periodicals would immediately become obsolete if the theory collapsed.

Although there seems to be enough evidence and clear logic to dump both the Big Bang and the expanding universe theories, it will probably take years to get the best minds in the field back on a productive track because of the panicky salvaging attitude of trying to stuff all the damaging observations and ideas into a continuously modified and patched model of the Big Bang. No matter how bizarre these necessary modifications become, it seems that anything is better than questioning the validity of the whole concept. The scientific community has dug itself and the rest of humanity into an intellectual hole from which there seems to be no easy or honorable escape.

One might logically ask why I am going to the trouble of recording these speculative theories, since my amateur status and lack of any observational or other hard evidence will surely minimize any consideration of the material by professionals. The experts in cosmology and particle physics are probably overwhelmed by the volume of incoming material just among their peers, so it is perfectly understandable that they would have very little time to spend on the wild speculations of an enthusiastic novice. Admittedly, this scavenger hunt for pearls of reality has mushroomed into a project that cannot possibly be comprehensively discussed by a single non-professional. Such a superficial treatment of this diversity of topics will undoubtedly be considered as a very unprofessional approach. Personally, however, I would feel sufficiently rewarded if this illumination of cracks in the body of standard theory plus the descriptions of logical alternatives operated as investigative stimuli for others.

Of course, there is only a very small probability for the emergence in the near future of any conclusive evidence bearing on such elusive and investigatively difficult subjects as the nature of the space entity or the true picture of subatomic structure and behavior. However, possibly in the year 2195, someone will establish the unity or the singular identity of the basic building block common to both the space and the mass phases of the universe. Then, one of my descendants might wave this piece of yellowing parchment around proclaiming that his ancestor had predicted this discovery 200 years before. Of course, there would be only a small amount of solace in the prospect for this type of posthumous notoriety.

Who knows, possibly we could get lucky in the next few years and detect observational evidence of expansion or dissolution at the galactic or quasar level, (a much slower or time expanded version of a super nova on the stellar level) thus completing an evidentiary circle that would suggest the probability of a very old chemical and structural universal equilibrium maintained by repeated cycles of formation, dissolution and reorganization of structures at all hierarchy size levels.

To me, this type of cosmology speculation continues to be a pleasant and rewarding endeavor in spite of the obvious and understandable difficulty I will continue to have in trying to bring these ideas to the attention of the busy professionals in the field. There is a sense of excitement produced by the prospect of opening closed doors and possibly discovering long hidden truths about our world that will probably continue to fuel my enthusiasm in these matters as long as I am lucky enough to stay one step ahead of senility.

With the exception of Hoyle and a few others, it has been a long time since the leaders in the field have been motivated or had reason to spend any significant intellectual energy on alternatives to the standard model of the universe.

SIZE, AGE AND MOTION OF THE UNIVERSE

Before getting into specific arguments about the size and age of the universe, we might spend some discussion time on the concept of infinity. Associative human thinking seems to require at least an abstract type of containment or boundary delineation of the phenomenon being considered in order for the thinker to feel comfortable in what he is doing. So, the contemplation of infinity tends to leave one with a sense of frustration and incompleteness that most would simply prefer to avoid. Apparently, having mentally evolved in an earth environment where containment type understanding is closely linked to survival, we find it difficult to shift gears and contemplate infinity without having an uneasy feeling of being on the wrong track. However, when considering the size or volume of the universe, the only alternative to infinity would be a finite universe, (attractive at first glance) which is actually much more of a logical impossibility than an infinite one. Recall that even the idea of limitless empty space around a finite universe leaves the concept in a state of open ended unresolve in that space must then be thought of as infinite.

Apparently, we must learn to accept the concept of infinity as being applicable to some aspect of reality, either the mass or the space phase of our universe. I would suggest that it is logically just as easy to let the idea of infinity apply to the extent of our currently observed type of universe as it is to claim that empty space must be the entity that populates this open ended volume.

In the realm of non-Euclidian universe geometries, Einstein's concept of a curved space-time continuum is occasionally used to model a criticism of the infinite volume universe idea. Discussing the credibility or reality status of this curved space-time idea is not within the scope of this manuscript. It is only mentioned here to highlight its mistaken use as follows. As a point against infinite size, it is sometimes argued that any prolonged journey or projection would eventually result in return of the traveling entity to the point of origin because of the curved nature of space-time. This argument, however, is a somewhat naive error in the application of the curved space-time idea. More correctly, it should be seen that the curvature concept implies that there should be curvature in all dimensions and all aspects of the space-time entity. There would be curvature of curvature or compounding of each dimensional curvature a la hierarchies of curvature. Such phrases as "change in the rate of the rate of the rate of curvature etc." would apply. The result is that a trip through curved space-time would have infinite directional diversity, and it would encounter an infinite series of locations with never a return to the point of origin or a duplication of locations even if the trip was of infinite duration. Thus, curvature of space-time and a universe of infinite size should not be thought of as mutually exclusive concepts.

Incidentally, some of the higher math forms like Riemann math and tensors might be pictured as attempts to incorporate the principle of compounding curvature into various geometries.

The three items listed in the chapter title (motion, size and age) make a pretty good outline of things that should be characterized in trying to develop a common understanding of the universe word. The possible variations among these three features can be summarized as follows. Size can be either finite or infinite. Age also has the same two possibilities of being either finite or infinite. Motion, however, can be described as expanding, gravitationally contracting or static. Obviously, with this number of variables, there are quite a few possible combinations. My objective will be to reduce this to one set of characterizations by eliminating the improbable variants.

Concerning the motion behavior of the universe, the overall static model will prevail as we eliminate the other two possibilities. The arguments for dropping the expanding model are sufficiently detailed in the introduction and my 1993 writing, so they will not be repeated at this point. The gravitationally contracting model is untenable because it carries the implication that everything around us would appear to be blue shifted which is not the case. Also, the implicit collapse time would not allow objects to have their currently determined ages of billions of years. Thus we have an overall static universe as the logical probability by elimination of the other two motion possibilities.

Next we will consider the size question (finite or infinite) and use some of the argument in the previous paragraphs to minimize the probability of a finite universe.

Arguing against a universe of finite size is something of an uphill battle because it seems that the weight of "common sense" has always been on the side of a limited universe. Historically, there has been a series of limitations placed on the size of the universe, all of them having in common the short-sighted reasoning that the universe cannot possibly be larger than the current range of vision or detection.

Possibly the most damaging argument against the finite universe concept concerns gravitational contraction. A finite universe cannot remain static. It must be in the process of an accelerating gravitational collapse. The rate of collapse and the total time for collapse are functions of the density and size of the universe. There is an inverse relation between the size of the universe and the rate of collapse. As the size of the universe approaches infinity, the rate of collapse (and our ability to detect that collapse) approaches zero. Thus, at infinite size, there is no possibility of overall contraction.

These same relationships would apply equally well to the concept of an expanding universe, which means; the larger the universe , the slower the rate of expansion and that the idea of expansion of an infinitely large universe would be an impossibility — a meaningless contradiction.

That probably covers the subject of the relations and contradictions between the size and motion parameters of universe models well enough for most readers, but I will agonize the details of the subject a little more in the following paragraph for those who might be interested.

Gravitational collapse implies directional preference in which everything is collapsing toward a single central point, but in an infinite universe there are an infinite number of locations, all of which would claim equal status as the center onto which gravitational concentration should impinge. Similarly, a boundary or periphery is a defining characteristic needed to give directional integrity to the concept of everything moving toward a center. An infinitely large universe has no such boundary, hence no particular direction that could be called a collapse direction. The standard Big Bang theorists are stuck with a finite universe no matter how improbable it may become. Their expanding model (whether it is "open" or "closed" concerning its eventual behavior makes no difference) must necessarily presume that the total mass of the universe is contained within a finite sphere some 20-30 billion light years in diameter with nothing beyond that boundary. Any attempt to deny that this expanding universe has such a periphery or outer limit would necessarily imply a boundless or infinitely large universe which is conceptually contradictory to the idea of expansion.

Recall that the background microwave radiation has always been explained in the standard model as a vestige of the activity in the early seconds of the Big Bang. When the long overdue demise of the standard model finally arrives, this microwave radiation will resurface as an extremely important piece of determining evidence concerning both the size and age of the universe. Everything about the background radiation screams out for the obvious explanation — that it is highly red shifted stellar/galactic radiation from very numerous and very far away galactic sources.

During the 30 years since background discovery, this glaring high probability explanation has received practically no serious consideration or discussion in cosmology circles because it is completely contradictory to the Big Bang — expanding universe model. Typically, the bulk of intellectual energy has been directed along the lines of trying to cram the mechanism of this radiation event into a contrived band aid type of early Big Bang modification that could explain the manufacture and present day existence of the smooth microwave background.

The observed features of the background radiation logically and clearly point toward its origin. Equal intensity from all directions shows that the radiation comes not from our galaxy (because of our eccentric location), but from the stellar fusion activity of more distant uniformly distributed galaxies. The smoothness would indicate a very tiny projection angle between galactic sources which means that the radiation originates from thousands of times more galaxies located thousands of times more distant than our currently recognized population of galaxies. The average microwave wave length is consistent with galactic radiation that has been highly red shifted (by a factor of $z = 100$ to 300 or more). This obviously implies that the red shift is a function of the distance rather than having anything to do with recession velocity or expansion. The possible mechanism for distance caused red shift is covered in the universal entity chapter. The above logical associations clearly suggest a universe size that is at least thousands of times greater in diameter than the 20-30 billion light year limit imposed by the standard Big Bang model.

The concept of background radiation as being highly red shifted radiation from distant galaxies also carries rather profound implications concerning the minimum age of the universe. In the currently detectable portion of the universe, a red shift of 4 represents energy coming from a distance of about 1.5×10^{10} light years which means about 1.5×10^{10} years in transit. The average wave length of the background radiation suggests that it has been distance red shifted more (by a factor of about 100 or so) than the maximum red shift distance combinations currently accepted; so this radiation must have begun its journey about 1.5×10^{12} years ago.

Who wants to be first to jump up and suggest that the age of the universe must therefore be about 1.5×10^{12} years, and that its edge is probably located about 1.5 trillion light years from here? Not me. I would, however, suggest that these ideas, rather than logically limiting the theoretical size and age of the universe, tend to raise the probability status of a universe that is infinite both in size and age.

LIGHT TRANSMISSION

In this section the terms "light" and "energy" will be used interchangeably. To facilitate the narrative flow, light will also be used in a generic sense to mean any part of the electromagnetic wave spectrum from gamma radiation to visible light to radio waves. Any spherical wave disturbance that propagates in a vacuum with a speed of "c" (300,000 Km./sec.) will be referred to as light.

Even today there is no general agreement concerning the definition of the phenomenon called light. The arguments range back and forth between various combinations of a media dependent spherical wave disturbance versus models of a projectile or ray-like penetration by an entity described as a particle, packet or quantum. The reality of the situation will someday emerge and it will be very close to one of the two divergent concepts noted above. It probably will not, for example, settle down to a vague duplicitous model like quantum mechanics with its uncertainty principle that is even now widely accepted in explaining sub atomic behavior.

My opinion is that light is not a structural entity of any type. It is simply an event or phenomenon that is best described as a disturbance. This definition makes it necessary for the propagating space medium to have some kind of structure and behavior pattern that is capable of being disturbed. The concept of "empty" space, for example, would necessarily deny the existence of anything within this light transmitting volume that could be disturbed.

The fact that the known speed of light in such a huge

number when compared to the velocity of other types of wave disturbances is a clue that should not be ignored. It suggests that the permanent, natural state of high speed curved motion that almost certainly characterizes this light transmitting space entity must also be quantatively vastly different in its parameters from the behavior of media units like air molecules in the wave transmission of disturbances such as sound. However, I think sound offers us some valuable and valid qualitative comparisons in trying to understand light transmission. The spherical spread pattern of light closely resembles known terrestrial disturbance or vibration patterns like sound, shock waves, and pebbles in a pond.

Historically, the speed of light was very naturally thought to be infinite or instantaneous until about 300 years ago when O. Roemer first demonstrated its finite velocity, which has since been more accurately defined through the years to reach today's highly precise value of c.

During and before the 1700s, the structure of light was considered to be a stream of tiny particles, which seemed to fit well enough with its behavior concerning reflection, sharp shadows, and finite velocity. Newton had favored the particle theory of light mainly because it cast sharp shadows.

Starting at about 1800, the wave theory of light began to take over because of the work and observations of a number of enthusiastic scientists. Thomas Young, in 1800, cited the observed interference phenomenon as important evidence for the wave mechanism. From 1815 to 1850, Fresnel and others strongly reinforced the wave concept by their work on refraction and variable speed of light in different media. Polarization experiments created a temporary paradox at about this time. This was satisfactorily resolved, however, by changing the theoretical direction of light wave oscillation from longitudinal to horizontal.

During these years, Maxwell and others realized that a medium (or ether as it came to be called) was a complete necessity for this wave transmission of light. In the late 1800s there were quite a number of theories proposed by an equal number of energetic scientists on the structure and behavior of the space medium or ether. I get the impression that this was a highly competitive era in which scientists may have been overly ambitious in their zeal to court public favor for their pet ether theories. Some of these ether models succumbed to observational evidence, so that in about 1895 the consensus seemed to be one in which the ether was seen as an all pervasive medium that extended throughout all space and mass. The ether was not considered to have any structural or behavioral similarity to mass with which it shared a common space or volume. It was also presumed that gravity or the universal cohesive force had no effect on the ether. The most critical and unfounded assumption about the nature of the ether concerned its lack of motion. All mass responds to gravity by contraction and a myriad of rotational and revolutionary motions while the ether remains universally static, just a co-existing stable background or matrix for the superimposed actively gyrating mass objects. The ether was thought of as being physically unconnected and qualitatively different from the mass with which it shared a common volume. The inclusion of this somewhat metaphysical "static" idea in the ether scenario would eventually cause problems for the theoretical existence of a space medium for light propagation.

In the last few years of the 19th century there were a number of similar light velocity experiments which can be conveniently grouped as the Michelson-Morley experiments. The essence of these experiments was the surprising observation that there was no difference in the speed of light (no variation of c from its standard value) because of the earth's movement in respect to a static ether. The fact that an earth observer was approaching or receding from a distant light source had no effect on c as measured by that earth observer. This created a paradox because it was assumed that the velocity of light should always be a constant in its preferred medium — the static ether. The movement of the earth observer through this static medium should have altered his measurement of c which did not occur. The response of the scientific community to this paradox was generally one of doubting or denying the existence of an ether. More realistically they might have reasoned that the "static" assumption about the nature of the ether was the only problem in this conflicting situation. However, overreaction seemed to prevail, so they threw out the baby with the bath water by denying the existence of a light transmitting ether.

Near the beginning of his famous 1905 paper on special relatively, Einstein stated that "the introduction of a 'luminiferous ether' will prove to be superfluous." This dismissal of the ether was a prelude to (or baseline assumption that required) the relativistic alterations of time and length which would appear later in the article. The leading theorists around the turn of the century seemed to be unwilling to recognize the possibility that relative velocity might exist between contiguous volumes of the light transmitting medium. This would have been a very tame assumption — just making light analogous to sound concerning the concept of relative velocity between different volumes of media, since this concept was already an accepted feature of sound transmission in air. Furthermore, the relative motion capability would suggest that the ether or space medium is responsive to gravity, since we already accept the comparable situation in which relative motion between neighboring mass volumes is a gravity driven activity.

Most would agree that a structured medium or ether is a necessity in explaining a wave like propagation of light through space, thus it would have been more logical and less drastic for Einstein and others to have reasoned that the assumed behavior of the ether (rather than the existence of the ether) was causing the Michelson-Morley observational paradox. Of course, accepting these space behavior concepts would also have made the much more bizarre assumptions of special relativity completely unnecessary.

Around 1900 there was some renewed opposition to the wave theory because of observations suggesting that the particle or corpuscular concept of light could explain some things better than the wave mechanism. The theory holds that light may propagate in a corpuscular manner as discrete packets of energy. There are a number of experiments that suggest this segmental behavior. A frequently cited example is the observation that light impacting mass causes an energy exchange in which the energy departing the collision site seems to do so in discrete amounts called quanta. At the moment, I don't have the knowledge or the time to give a properly detailed discussion of this apparent corpuscular behavior of energy, so I will leave the subject with the following comparative opinion. There are multiple observations and logical arguments that make a very strong case for the identification of light as a media dependent spherical wave disturbance. It is my opinion that the quantum features of light will eventually be understood more fully and subsequently integrated into the wave concept rather than serving as an instrument to overturn that well documented reality.

This is a convenient place to review the development of the logic concerning a light transmitting space medium by using sound in an analogy. Even though the quantitative parameters of sound and light are vastly different, there are still valuable comparisons that can be made between the two wave mechanisms. Both sound and light are disturbances (rather than structural entities) that pass through a medium in a spherical wave like fashion. I would further contend that the propagating velocity in both cases is completely dependent on the structure, behavior, and associative geometry of the units that make up their respective media.

Concerning sound in air, this idea of media dependent velocity means that the shape, size, texture, and spatial proximity of individual gas molecules is a pertinent factor. The other factor of importance concerns the natural state of movement or vibratory activity that is characteristic of these particular molecules. This statement applies to any movement of the entire molecule if it exists, but mainly it applies to the hierarchy of movement that is inside the periphery of the molecule. The parameters of this natural, permanent, internal movement determine the method and speed of sound transmission across the diameter of a molecule when a sound impulse disturbs or indents one side of the molecule. So, sound transmission in air is a function of both the internal and external motion behavior as well as the internal and external architecture of the gas molecules that compose the medium.

Let's picture a narrative model in which a stationary source is emitting a continuous sound into the surrounding air. The sound has a wave length of 1 meter and its leading edge propagates through air at 340 m/sec. There is a physics lab within a vehicle that is moving away from the source at 1/2 the speed of sound or 170 m/sec. The vehicle is closed and contains air, so we have a situation in which there is relative velocity between two contiguous volumes of sound transmitting medium. When a sound impulse from the source penetrates the rear of the vehicle, it instantaneously shifts gears and assumes the speed of 340 m/sec. in the contained air of the vehicle. The instruments at the center of the vehicle record this sound speed of 340 m/sec. and they also record that the wave length has changed to 1.5 m. So, the lab personnel could easily determine their own ground speed by knowing the speed of sound in air and the percentage change in the sound wave lengths that enter the rear of the vehicle. An outside observer would see a deformity in the shape of the wave front as it passed over and through the vehicle. The portion of the wave that penetrated the rear of the vehicle would appear to bulge forward and speed up relative to the other portion of the wave outside the vehicle. In fact, the outside observer would measure the ground speed of the portion of the wave that was inside the vehicle as being 510 m/sec. — the speed of sound in air plus the speed of the vehicle.

Instruments attached to the outside of the vehicle would tell a different story from the inside ones. There would be no change in the 1 m. wave length, but the sound waves would pass these instruments with a frequency of 170/sec. and a speed reduced to 170 m./sec. Let's keep in mind that this reduced speed reading occurs because the recipient of the wave impulses is in motion relative to the sound medium. It would not have occurred as a result of the source being in motion relative to the medium. In other words, the fact that there is a single value to describe the separation velocity between source and recipient does not mean that the model is symmetrical regarding interchangeability of source and recipient. Of course, the media dependence of the speed of sound is the reason for this asymmetry. However, it could be made symmetrical and interchangeable if the source and recipient were separating in such a way that each had the same velocity relative to the medium.

Possibly I have belabored and over explained this sound model. Obviously, I am trying to make the point that this source — recipient — velocity — wave length — medium scenario has some applicability to the light transmission situation. In fact, if we drop the relativity assumptions, including the speed limit of c, and if we assume the possibility of relative velocity between contiguous volumes of light transmitting media, then light would behave very much like sound and would also accommodate the observations of the Michelson-Morley experiments.

In my preferred model of a single universal entity, the space phase of this entity is physically continuous throughout the mass phase just like the old ether idea. However, this model is different in that all parts of the universal entity are mobile and all are affected by gravity. The parts of the entity that are superimposed throughout a mass structure are gravitationally held into a cohesive unit just like the atmosphere adheres to the more dense earth in its rotational movement. So the whole cohesive complex (earth, atmosphere, pervasive and surrounding shell of ether or space medium, and even the Michelson-Morley observer) is gravitationally bound together and rotates with the earth as a unit.

The following is a short digression to clarify what is meant by the recession of the Michelson-Morley observer (the earth bound lab that measures the speed of light) relative to a distant light source. Let's say that, relative to this distant source, the Michelson-Morley observer is located on the side of the rotating planet that is receding from the source, and that the earth is similarly located in the portion of its solar orbit that is also receding from the distant stellar source. Thus the Michelson-Morley observer's recession from the source will be the sum of these two motions. This recession total would be some several hundred km./sec. which I will not bother to define more accurately at this time. Therefore, this recession total should be implied whenever the rotational speed or recession of the earth bound Michelson-Morley observer is mentioned.

With the above model in mind, let's re-examine the behavior of light from a distant stellar source as it approaches earth. The earth is rotating and revolving so that the Michelson-Morley observer (or light physics lab) is receding from the distant light source. As the light wave enters the shell of space medium adherent to earth, it instantaneously shifts gears and assumes the velocity of c within its new medium. The Michelson-Morley observer on earth records the usual unchanged value of c, and he would also record a tiny wave length increase or red shift of the incoming light if his instruments were capable of detecting such a necessarily tiny percentage change in wave length. Let's also presume that there is a second light measuring observer located some distance away from the solar system in a volume of the inter-stellar medium that is static relative to the distant stellar light source. This location is physically separate from the shell of medium that adheres to the earth on its daily rotation, and it is also physically separate from the disc of rotating medium that accompanies all the planets in their solar orbits. The second observer is self propelled and maintains a position that is static relative to the earth observer, as if they were tethered together. This means that the second observer mimics the earth observer's rotational and solar orbital movement so that both observers are at all times located the same distance from the distant stellar light source. In these circumstances, the second observer would record the same standard wave length that was determined by the input frequency at the distant source (no red shift), but he would also record the incoming stellar light as having a velocity less than c by an amount equal to the sum of the rotational and solar orbital speeds of earth. This is true because the second observer is a recipient who has motion relative to the inter-stellar medium within which he resides. Therefore, this situation is analogous to the sound transmission model where the instruments were attached to the outside of the vehicle.

As a matter of experimental practicality, it comes to mind that a NASA type of earth orbiting satellite might be able to make measurements that would confirm the existence of both the cohesive shell of space medium that accompanies earth's daily rotation as well as the one that accompanies circum-solar planetary orbits. This satellite would make wave length, frequency, and light speed determinations of light from a distant source, which would be compared with similar observations on earth. Also, a satellite orbiting earth in the direction of earth's rotation might record light parameter figures that differ from those of a satellite orbiting in the opposite direction. Such an experiment might confirm some of the predictions of differences in light parameters suggested in the previously described model using two tethered types of Michelson-Morley observers.

The preceding sound and light models clearly portray my opinion of similarity between the two processes. It is also obvious that the necessary assumptions are quite tame when compared to the daring and unconventional presumptions that are required by special relativity.

THE UNIVERSAL ENTITY

Historically, the universe has been seen as a volume that has mass as its single widely dispersed item of content. Since the space between mass objects contains nothing that can be identified with even the smallest subdivision of mass particles, it was considered to be empty and completely devoid of any structure or substance.

Of course, in early times, gas and atmosphere were thought of as belonging to the empty portion of the universe volume. Then, with the advent of molecular science, the line of demarcation was adjusted so that air would be part of the mass component of the universe.

This obvious or common sense overview was not seriously challenged until several centuries ago when unexpected discoveries about light transmission through space began to accumulate. Evidence about the velocity and wave nature of light strongly suggested that space was behaving as a conducting medium and therefore should have some kind of structural integrity rather than being thought of as empty. Of course, the argument concerning structured or empty space is not yet resolved. I will discuss the role that light has played in this back and forth controversy about the nature of space in more depth a little later.

Before discussing the characterization of the content of the universe, I wish to make a retraction or modification of the picture I painted in my 1993 writing when I likened the activity of the universe to a triangular reversible equation in which mass, space and energy appeared to be qualitatively comparable entities occupying the three points of the triangle. This arrangement suggested a universe concept in which mass, space and energy represent a system of three interconnected sinks or storage bins among which the bulk of the universal entity maintains an equilibrium by passing freely back and forth between them. The analogy of equilibrium between storage bins might appropriately be used if space and mass were the only two participants, but reality distortion seems to occur when energy is thrown into the picture on an equal footing with the other two.

It would have been more correct to depict space and mass as the two structural entities that occupy the entire volume of the universe; then energy would be better envisioned as a behavioral reality or an eternal type of activity — a multi-directional type of disturbance acting through and throughout the two perceptual variants of the structural content of the universe.

Getting back to characterizing the texture of the universe, I would suggest that there are two basic arguments to be considered. One says that space and mass are two completely different entities, separately derived and existing side by side in the universe without overlapping or confluence. The other one suggests that space and mass are just two perceptual variants of a single entity.

Common sense and going along with the obvious would seem to favor the two entity universe. This concept is reinforced by the stark contrast or sharpness of delineation between space and mass as we observe it today. For example, we have the earth sitting in a nest of space, a situation that has existed for billions of years with no apparent mixing or transition gradient at the relatively sharp demarcation between the two.

Having painted myself into a corner with such a compelling picture of a universe containing two entities, I would still like to present the argument for a single entity universe; one in which the extreme contrast between mass and space is seen as a case of mixing (for billions of years) resulting in the random juxta positioning of two widely separated members in the gravitationally driven spectrum of the structural change of space. Think of it as a very slow spectrum of concentration or densification that is continuously converting the space phase entity into mass.

The idea that gravity alters or concentrates space entity sufficiently to make it pass the threshold of human mass perception (at about proton size) becomes easier to philosophically accept when it is compared to what we already accept about fusion and evolution. Some protons or hydrogen atoms, driven solely by this universal cohesive force, will undergo fusion, expansion, re-contraction, and finally evolution to become biologic forms. In other words, the continuous gravitational manufacture of protons from the space phase of the entity is much less of a conceptual pill to swallow than the natural gravity driven process of hydrogen to humans that is already accepted. This space to mass process would be occurring very slowly and continuously in all intra and intergalactic space and would, of course, be undetectable to us until volumes of hydrogen (intra galactic gas clouds?) randomly appeared.

In an infinite or very old universe, this production of mass from space would clearly be necessary to maintain present day equilibrium. The disappearance of mass accompanying stellar fusion at something near the rate of 6 million tons per second (for the sun) would not allow current equilibrium without some way of converting space to mass. The equilibrium situation will be covered in more detail a little later.

Let's go back for a moment to the stark contrast in appearance presented by massive objects like stars and planets on a background of space. The physical proximity of these two entities does not imply that space somehow transforms directly to a planetary or stellar mass form. The contrast is caused by an entropy type of mixing resulting from various natural expansion processes like red giants, novae, super novae, and even galaxy or quasar expansions, etc. These expansions might be thought of as a type of sudden retrograde activity (precipitous fusion thresholds are probably important in this behavior) when compared to the smoother gradient process of concentrating space into mass. Both processes are occurring continuously in the same universe and we should not be misled by our vivid visual awareness of the results of retrograde mixing, or spectrum convolutions, as compared to our lack of ability to perceive the space to mass process.

I would suggest that mass and space are just two different human perceptions of the gravity driven structural and behavioral variants of a single entity that is eternally involved in a dynamic equilibrium between its space and mass phases.

To the extent that I may have been successful in selling the probability of a single entity universe to the reader, let's try to characterize the space phase of this entity. There is obviously no basis for saying anything specific about the structure of this elusive entity. The best we might anticipate would be the prospect of gaining some insight through deduced or observed behavior patterns.

First, concerning the distribution of this space entity, there is evidence for the continuous uninterrupted extension of this entity throughout the fabric of all mass as well as space volume. This concept is reinforced by the similarity in the way light is transmitted through both space and mass entities such as air and glass. The velocity of light through air and glass is only fractionally reduced from its value in a vacuum. For example, the value of "c" in air is 99.97 percent of the value in a vacuum, which clearly shows that the light disturbance is vibrating through the space medium texture that must exist throughout the same volume occupied by air molecules. There is no conceivable way that air molecules could be directly involved in this extremely rapid propagation. Remember that sound uses the naturally occurring motion harmonics and elasticity features of adjacent air molecules to propagate its disturbance at 343 m/sec., a rate that is about a million times slower than the speed of light through air. Glass transmits light at about 200,000 km/sec. which is approximately one-third slower than the vacuum speed. The speed of light through water is between that of glass and air. So it should be obvious that these fractional variations of c in air, water and glass mean that the density of the space entity that occupies the same volume as the mass lattice work is being increased fractionally by the cohesive (or gravitational) contracting influence of this surrounding lattice. The particular geometries of air, water and glass molecular organization are obviously such that they do not block visible light propagation through the co-existing space medium, but only serve to fractionally concentrate its structure which in turn fractionally reduces the speed of a traversing light disturbance.

My reference to the space medium undergoing concentration or increase in density is a generalized type of assumption that needs a little more clarification. Probably it would be closer to reality to say that glass or water alters the structural and behavioral geometry of its intersticial space medium in such a way that it resembles concentration or increase in density. However, as a matter of narrative convenience, I will continue to use the terms "concentration" or "density increase" in describing this medium alteration.

In a philosophical overview, it would not seem possible for particulate or corpuscular light to maintain an exact speed of c for billions of years unless the energy determinant of this speed was a built in natural feature of the medium. A medium with a natural, permanent, baseline behavior of high speed curved or rotational motion would much more logically be capable of propagating a disturbance (rather than quantum packet) for billions of years with no change in speed.

Let's consider another analogy in which a light beam is directed through a series of separated glass blocks. As light enters the first block, its speed drops to 67 percent of c, then increases again to almost c in the air space when it exits the first block, then down to 67 percent of c again as it enters the second block and so on. This oscillation strongly suggests that the speed of light is media determined. I cannot envision a mechanism whereby corpuscular light through empty space (empty space that it interstitially pervasive throughout the air and glass volumes) could behave in this manner.

Since space medium and mass are variants of a single entity, it is logical to presume that the universal cohesive force (gravity) acts on both. As a baseline for comparison, I would first like to characterize the cohesive behavior of earth for its atmosphere in a non-mathematical fashion. The attractive aspect of gravity causes the atmosphere to be dragged along, sticking to the earth in its rotational motion rather than allowing the earth to slide beneath a stationary shell of atmosphere. The reason for this is that the cohesive force between earth and atmosphere is greater than the cohesion between atmosphere and its adjoining less dense space medium. Therefore, the sliding or gradient of differential motion between the rotating earth and the non-rotating volume of the solar system disc occurs in the space outside the earth's atmosphere.

The other easily recognized qualitative action of gravity is the volumetric reduction or concentration of mass even at the atomic level. The gradient of this density increase is admittedly slow or gradual, (It takes a large quantity of mass to furnish the pressure needed to reduce an atom of iron to half its volume.) but the tendency exists, so it must be presumed that the space phase of the entity surrounding a massive object like the sun must also have a density gradient that decreases moving away from the object. More will be made of this later in considering the mechanism of light transmission alteration in the vicinity of the sun.

Consider the following thought experiment as possibly helpful in understanding the motion patterns between different phases of the universal entity. Presume that universal cohesion causes the self attraction of all parts of this single entity universe for all other parts and that the strength of the attraction or cohesion between areas is proportional to the product of the density factors of any two areas. Then, arbitrarily assigning a density factor of 100 to mass and 5 to the space phase of the entity, we have a situation in which two adjacent areas of the space medium would attract each other with a cohesive force rating of 5 × 5 or 25. Mass would cohere with adjoining space entity with a cohesive force rating of 5 × 100 or 500 and mass for mass would be 100 × 100 or 10,000. These numerical assignments of density and force relations between space and mass are admittedly nowhere near reality, but they should serve the purpose of determining the location where you would logically expect to see the necessary sliding or motion differential occur when you have rotation of mass adjacent to a more static space entity similar to the earth and its atmosphere rotating within the disc-shaped solar system volume of space phase entity.

To repeat for emphasis, the space phase of the universal entity is not confined to the volumes we identify as space. It is continuous throughout the fabric of all mass forms no matter whether they are gas, iron, or super dense objects. The space entity is not physically connected with the lattice work of mass with which it shares a common volume; however, the density of the space medium that resides within a mass super structure or skeleton is apparently increased by the cohesive force effect of this closely contiguous mass. Both of the above statements about the all pervasive nature and the variable density of the space phase of the universal entity are given some support by my previous discussion concerning the nature of light transmission and its variable speeds in glass and air.

I realize that the characterization of relations concerning mass, space, and cohesion in the last few pages may seem complicated and difficult to follow, due partly to my inability to state it as clearly as it should be presented. However, I am trying to justify a picture in which the earth, its atmosphere, and the light transmitting space medium trapped within the matrix of the atmosphere are all three gravitationally adherent to each other and rotate as a unit. Beyond this rotating unit, maybe 3-4 earth diameters away, the space medium is more static in that it has practically no rotational motion matching that of the earth. The exact location, width, and gradient of this motion transition is unknown. The significant point to be demonstrated, however, is the relative motion between two contiguous volumes of light transmitting medium. The importance of this relative motion to the perception of the speed of light by various observers is covered in other chapters.

In trying to describe the motion of the space entity, since it behaves similarly to mass in response to the universal cohesive force, it could be said that it tends to go along with the overall movement or rotation of whatever massive object system within which it resides. Thus, the solar system represents a thin rotating disc of mainly space medium within which are imbedded mass objects of various sizes like planets, moons, asteroids, gas and dust, all rotating as a unit. So, all the components of this disc (space medium and various mass objects) are almost static relative to each other. This statement is not entirely correct, however, since there are obviously many smaller hierarchy levels of rotation within the solar system disc such as moons around planets, etc.

With this in mind, galaxies would seem to fall into place as the next larger hierarchy level of disc like rotation of the mass-space entity complex. The galaxy contains some 10^{11} stars (or rotating solar system like discs) all rotating with, and remaining generally static relative to, their intervening space medium volume at a speed of about 250 million years per revolution. Of course, there may even be a tendency for galaxies to form larger rotating discs of clusters or super clusters. However, as discussed by me in 1993, this tendency may be obscure because the margins of unit delineation seem to become progressively less distinctive as larger levels of hierarchy organization are observed.

This multi-level organization of rotating systems within rotating systems is significant in that it once again demonstrates the important point that there are large and varied motion differentials between contiguous volumes of the space entity.

The following analogy may help in visualizing motion patterns of the space phase entity within a galaxy. Picture a large auditorium or astrodome containing several dozen widely spaced slowly moving rotary fans. There is random variation in rotational direction and planar orientation of the fans. Each fan has multiple blades that are perpendicular to the plane of rotation. Also, the astrodome is filled with smoke to visualize the medium movement. The smoke between blades tends to move along with the blade motion, but beyond the blade tips there will be a rapid (steep gradient type) decrease in the tendency of the smoke to follow the blade motion. Yet there will still be some smoke movement extending from each fan to all parts of the astrodome. The movement induced from all fans will interact in the space between fans to form a continuing and infinitely complicated pattern of multi-directional, curved, and swirling motion. However, the movement of smoke between fans will be relatively slow and it will average out in such a way that the sum of the vectors of smoke movement between fans will be close to zero.

The fan — smoke — astrodome analogy is meant to show the motion patterns that result when gravity acts with a variable intensity because of a difference in density between the recipient mass forms. Therefore, some insight about space entity motion may be gained from the analogy to the extent that the space entity can be considered to be a less concentrated form of mass.

This is admittedly a very rough and inaccurate analogy, but it may serve to give the reader a feel for the behavior of volumes of space phase entity within a galaxy, and, of course, between galaxies for that matter.

This is an appropriate time to revisit the possible mechanisms for Hubble's red shifts — the mostly linear relationship that was found to exist between red shift and the estimated distance to moderately remote galaxies. Let's presume that arguments by me and others have sufficiently put the standard expanding universe model to rest so that we are free to seek a reason for this red shift that is unrelated to recession velocity of the galactic light source. This means that we should look to the structure and behavior of the many light years of intervening space medium for an explanation of the observed lengthening of wave length.

Let's also presume that the last couple of chapters have established two gravity driven features of this intergalactic space medium as follows. The space medium has motion between its various volumes that roughly parallels the motion patterns among mass accumulations in the universe. Also that there is a small range of natural gravitational density of the space medium that ranges from a baseline low in intergalactic space to a high in the vicinity of stars. Thus, the speed of light has a correspondingly small range of variability (inversely related to the density) from slightly above c in intergalactic space to slightly below c in the vicinity of stars.

I would suggest that it might be possible to use these two variable features of the space medium to theorize an increase of light wave length (red shift) that would be additive or cumulative in its nature to match the linear or additive nature of Hubble's distance red shift observations. Considering the media dependent nature of light, it would then be logical to assume that these two media features of variable density and variable motion would cause, respectively, small fluctuations in the speed and direction of traversing light waves. This effect might be pictured as causing a tiny segment of light wave to have a serpentine rather than a straight line path in its intergalactic travel. Another conceptual model would be to consider light waves as concentric spherical shells in which case the surface of these shells would be wavy or rough (rather than smooth) because of these two media feature variations. Furthermore, and most importantly, the wavy pattern on the surface of one wave shell would not match or mesh with the pattern on the surface of its contiguous or neighboring shell. This is true because there would be changes in both of these media features (variable density and motion between volumes) within the space between two contiguous wave shells that would occur in the short time interval between the passage of two consecutive waves.

Trying to trace the effects of media variability on wave length beyond this point becomes highly speculative and conceptually more difficult to describe. I will suggest just one possible sequence in regard to mechanisms of wave length increase. The deflections of light within each inter wave crest segment due to media variability might normally be considered to be evenly divided between those that would lengthen and those that would shorten the travel path of light, except for the fact that some of these deflections would cross the zero or straight ahead direction, in which case the lengthening deflections would outweigh the straightening ones, thus accounting for a progressive increase in the distance between wave crests (red shift). Admittedly, this is just a speculative sequence that may contain naive assumptive errors. It is included at this point only to demonstrate that there may be a number of mathematical models to consider in trying to account for such an extremely minute incremental change in wave length that is the only requirement for explaining the distance related red shifts. It should be remembered that red shift represents a cumulative wave length change that is almost unbelievably gradual. Light traveling a distance of about 6 billion light years (6×10^{25} m) causes a red shift of Z = 1 which might be spectrographically represented as a wave length change of some 10^{-6} m. This means that the ratio of distance traveled to wave length change is about 6×10^{31} to 1.

There is one experimental analogy using sound that comes to mind in trying to demonstrate how variable media movement might cause a cumulative lengthening of wave length. Fans could be used to agitate the air medium in a sound transmitting tube or tunnel. Measurements would be made to determine the effect of this agitation on sound wave length as well as any proportionality between this effect and the length of the tube.

So, in leaving this model, I would suggest that enough variability in media dependent light behavior has been established to possibly allow the development of a mathematical model that could link distance red shifts to media variables. I think it is reasonable to say that the distance related red shifts are almost certainly related to something that is happening in the intergalactic space medium rather than to recession velocity or expansion. This "something" might be related to the media features described above, or possibly to some yet to be discovered structural or behavioral features of this light transmitting space phase of the universal entity.

RELATIVITY, THE DRAMATIC MISADVENTURE

In the early 1900s, there was considerable confusion about the definition of the light phenomenon as well as its method of propagation. This uncertainty also included the structural and behavioral features of the space through which light moved. This should not suggest that we are even today free from confusion and uncertainty on these subjects.

Einstein was only one of a number of talented thinkers in that era who were struggling to define concepts of light and space that would accommodate contemporary observations, especially those concerning the speed of light. History has a way of attaching particular names to periods of radical change in scientific theory and so it was with Albert Einstein. Today, he personally receives most of the credit (or blame as the case may be) for all the predictions and assumptions within both the special and general theories of relativity. Actually there were a number of talented scientists like Maxwell, Lorentz, Poincaré, Fresnel, Fitzgerald, and others who contributed to the theoretical revolutions of that very active period. Since most of my speculations are generally skeptical and adversely critical of everything in both relativity theories, this historical focus on Einstein does nothing to advance the popular acceptance of my ideas.

I don't wish to appear to be picking on Einstein for two reasons; first, because I have the highest personal respect for a genius of such high integrity. Secondly, I am aware that readers are likely to be turned off if it seems that I am singling out the most beloved and respected scientist of recent times as a whipping boy. However, there is no way for me to completely avoid this unfortunate perception, since the theories and assumptions attributed to Einstein are so much at the core of the flawed conceptual logic that has resulted in today's skewed view of time and space reality.

Einstein came along at a time when the public and the scientific community seemed to be ready for such an intellectual role model and leader. His popularity was enhanced by the perception that he was a self made mathematical genius from a humble background who had always been a fearless, imaginative, and unconventional ground breaker. He was not easily discouraged and seemed to have confidence in the belief that he had the unique ability to put it all together. To me, it seems clear that this public and scientific community perception of such a personable and energetic young wizard must have made the acceptance of these bizarre theoretical concepts occur with more ease and less skeptical resistance than might have been the case with a less charismatic lobbyist.

Einstein seemed to have a talent for introducing the nidus of new ideas in a benign and understated fashion that left the door open for his peers to extrapolate them into new and boldly revolutionary rules for behavior and structure. For instance, in his famous 1905 paper on relativity, he began with two postulates which I will briefly paraphrase. Number 1, called the principle of relativity, says that the unaccelerated motion of a vehicle has no effect on the way things happen within that vehicle, which is another way of saying that the laws of physics are the same within all unaccelerated reference frames. This seemed to be just an innocuous reaffirmation of behavior that was already widely accepted.

The number 2 principle in the 1905 paper said that the motion of light is not affected by the motion of the source of that light. This, however, was a somewhat vague and incomplete statement in that it did not specify source motion as being relative to either the recipient, the transmitting medium, or both. Actually, this second principle causes no conceptual problems if it is accepted that the transmission speed of light is completely determined by the texture of the medium within which it is traveling.

Before getting into a critical analyses of special relatively, I wish to repeat and elaborate a little more on my preferred concept of light as a media dependent disturbance as follows.

A thought model for this concept contains three essential elements — the light source, the light recipient, and the intervening and surrounding transmission medium (or space medium). The only behavior rule for light is that this spherical wave disturbance must always propagate through the structure of space medium with a constant speed value of c. Next comes the complicating part. Each of these three basic elements (source, recipient, and medium) is capable of multi-directional movement relative to either one or both of the other two. Obviously this compounding of variables will allow a wide range of "perceived" light velocities to occur within this one frame of reference. To aid in understanding the situation, I will describe a few representative examples of the resulting motion patterns.

In the first case, the recipient and medium are at rest relative to each other and the source is moving away from these two; thus the emitted light is dealt into the medium with an increase in wave length, reaching the recipient with a decreased frequency and the standard speed of c.

In the next case, the source and medium are at relative rest and the recipient is moving away; so the light pulses enter the medium with the standard wave length and they are received by the recipient with a decrease in both frequency and c.

In the third situation, source and recipient are at relative rest and the medium between the two is moving away from the recipient and toward the source. The result is that the wave length distance is spatially compressed (shortened) as it enters the medium and the recipient records a decrease in c but no change in frequency.

These examples demonstrate that wave length is a linear distance quantity that is determined at the source by the relative motion between source and medium. The recipient can arbitrarily change the value of both c and frequency by his variable motion (relative to the medium) but not the wave length. It should also be recalled that this entire behavior scenario is very closely analogous to the sound transmission model that was constructed in a previous chapter.

One other item to keep in mind is that this media dependent model is not symmetrical in that source and recipient are not freely interchangeable. The importance of the media would force a change in the values of the model if the recipient and the source changed places, as a review of the case examples demonstrates.

To avoid grossly over explaining the situation, I will simply state that — by keeping the behavior rule about media light speed in mind, the reader should be able to compare and agonize these three examples into an adequate concept of what is meant by saying that light is a media dependent phenomenon.

Now, getting back to the early 1900s, Einstein and his contemporaries seemed reluctant to concede the previously described variability of c (essential to the media dependent model), probably because of their misinterpretation of the information gained from the Michelson-Morley observations. — I am presuming that Einstein was aware of the Michelson-Morley class of experiments at this time. They had overlooked the probability that the constant values for c in that experiment were due to a situation in which the light transmitting space medium surrounding the observer was gravitationally bound to, and moving with, the earth and the observer so that the medium was static relative to that observer.

However, later in the 1905 paper we find the unfortunate statement that the idea of a medium or ether "would become superfluous." Einstein (and most of his contemporaries) seemed to be convinced that all observers must receive space transmitted light at the same speed, so the medium idea had to be dropped because it threatened to change the speed of light reception as previously explained. Actually, then, this statement about the ether becoming superfluous was something of a misleading understatement, because space would <u>have to be empty</u> in order to make the source and recipient interchangeable in a constant velocity situation as far as both recording the same value for c is concerned.

This denial of a medium set the stage for a dilemma in the following way. Without a medium, light must necessarily be considered as a corpuscular type of ray whose speed is determined at the source, so that it always separates from the source at a constant speed of c. This created a problem because the recipient would then have to receive light at varying speeds to accommodate the variable motion between source and receiver. This, of course, was not allowed according to previous observations and assumptions. The leading turn of the century theorists had assumed themselves into an apparently impossible situation. It was at this point that Einstein came to the rescue by arbitrarily changing time and length, thus mathematically relieving the paradox. This forced solution, however, was not consistent with reality in that it created other conflicting and paradoxical situations which will soon be demonstrated by way of a model and thought experiment.

The typical setting for demonstrating the provisions of the Special Theory of Relativity involves two frames of reference where there is unaccelerated motion between the two. This can be described a little more clearly as follows; one system, or reference frame, involves a light source and one or more light recipients or observers where there is no motion between source and recipient. The second system is composed of one or more recipients or observers that have steady or unaccelerated motion relative to the first system. In this set up, the special theory predicts that the observers in the second system will derive time and distance values to characterize relations between the two systems that are different from the time and distance values that the members of the first system will declare. Of course, these discrepancies are mandated by the assumption that the speed of light must always be perceived as c by all observers regardless of relative motion. This reasoning also must necessarily deny that the texture or movement of the medium has any effect on perceived light speed, which in effect is a denial of the existence of a light transmitting medium.

In the mathematical resolution or explanation of the special theory of relativity, there is one frequently used math term called the Lorentz contraction or the Lorentz transformation ($\sqrt{1 - v^2/c^2}$) that deserves special attention. The expression was derived by several scientists including Einstein and Lorentz. v refers to the velocity difference between two light systems (reference frames), and, of course, c is the speed of light. This Lorentz contraction (L.C.) is the factor that is used to modify both time and distance values in order to mathematically equate two motion different systems under special relativity assumptions. As long as v is less than the speed of light, this term (L.C.) is always less than 1, so it is commonly used to slightly reduce the value of any other term that it precedes in a modifying or factoring position. It can be seen that when v = 0, L.C. becomes 1 and has no effect on the term it precedes. It can also be seen that for everyday values of v, the ratio of v/c is a tiny number and that placing the square designations on the ratio (v^2/c^2) will greatly exaggerate the minuteness of that number. So, at everyday velocity values between two systems, the square aspect (v^2/c^2) of the L.C. term is of primary importance in causing it to operate as a very tiny reducer. The square root function of the L.C. term has only negligible effect at low or everyday values of v. It is only when v is almost as great as c that the relative importance of these two L.C. contained functions reverses.

Whether or not this L.C. term is the only, or even the most appropriate that could have been used in these system perception comparisons will not be pursued further at this time. Rather, I would prefer to argue that it is incorrect to use any reduction term to modify the perceived values of time or distance in relative motion models.

In the following model and thought experiment, we will first try to establish the verifiable reality of simultaneity and synchronization within one light system or reference frame where there is distance but no motion between members of the system. Later, a second system that has constant speed relative to the first will be added. Then, a series of simultaneous events will be constructed that are true and equally acceptable to all observers in both systems. And finally, the model will be used to show the contradictions and logical inconsistencies that flow from following the assumptions and perceptual dictates of special relativity.

Envision a model in which two physics labs (A and B) are located in a volume of intergalactic space that is far removed from large gravitational objects. The distance between

A and B is a measured and constant 1,800,000 km, or six light seconds. The space medium between A and B (if such a medium were conceptually allowed) would be static relative to both A and B. The labs at both locations have synchronized clocks that constantly send out light pulses at one second intervals. These pulses contain messages of time and date information so that when A's clock reads 100 sec. into January 25, 1995, he will be receiving the message from B at that same instant which reads 94 sec. into January 25, 1995. Of course, the reverse of this situation concerning B's clock readings is also true. Establishing this synchronization and absolute or proper time frame between A and B to the satisfaction of all observers will be analyzed a little more at this point.

The clocks at A and B can be used to establish or confirm the accuracy of the pre-measured distance of 1.8 million km between A and B in the following way; a light pulse sent from A to B and reflected by a mirror back to A will have a round trip time of 12 sec. as measured by A if the distance between A and B is 1.8 million km or 6 light sec. Also, the times required for light to span each leg of this round trip will be an equal 6 sec. in both cases. This is true because there is no light media movement relative to either A or B — a fact that can be confirmed by conducting 90° light interference (Michelson-Morley type) experiments at both A and B. The expected negative results in these experiments means that the speed of light is the same in both directions and that the 12 sec. duration of the reflected round trip is composed of two equal 6 sec. legs rather than a 5 and a 7 sec. leg for example, which could be the case if there was movement of the light transmitting medium relative to A and B. Thus we have established synchronized clocks and a confirmed distance between A and B as well as setting the stage for simultaneous events.

Any accidental aberrations in the relation values between

A and B would be detectable in the following ways. A change in the A-B distance would be manifest as a progressive and cumulative change in the 12 sec. round trip message time between A and B, while the date and time synchronization remained unchanged from day to day. Conversely, a malfunction in B's clock (running slow or fast) would be seen as a cumulative day by day mismatch in the original time and date synchronization between A and B, while the 12 sec. round trip pulse time remained unchanged. In the event that movement developed between the light medium and the AB observers, the above two values (12 sec. pulse round trip and time-date matching) would remain the same, but the results in the Michelson-Morley experiments at A and B would show changes. Lastly, in the case of clocks that were synchronized before separation, any temporary change in clock function due to the separation process would cease when the clocks reached their permanent points of separation at A and B, and it would also be detectable in the subsequent AB exchanges as a small aberration in the original time-date synchronization that would not change from day to day.

So, in summary, any changes in values pertaining to the relation between A and B, such as clock function, relative clock motion, media movement, or distance changes, would be detectable and identifiable as to their type and quantity by the particular signature left on the messages between A and B or on the results of the Michelson-Morley experiments conducted at both sites.

I would suggest that the concepts of synchronization and potential simultaneity of events between two distant but static observers should be accepted as verifiable realities. However, most written accounts on the subject reveal that relativists still tend to disparage or question the concepts of synchronization and simultaneity by artificially limiting the permissible deductive methods to only that information that can be gained by the exchange of current light speed messages between observers. *(For convenience, the term "relativists" will be used to designate those who adhere closely to the postulates and predictions of special relativity.)*

The next model component to be introduced is another fully equipped physics lab called TR (for Traveler). TR is self propelled and maintains a constant speed of 1/2 c (or 150,000 km/sec.) in his straight line trip from A to B. It is pre-arranged and understood by all synchronized labs that TR should pass through A at the instant designated as <u>time zero</u>. TR's clocks and instruments are synchronized with those at A by direct physical contact as TR passes through A. Then TR and B exchange all their recorded information at the instant TR passes through B. For example, this exchange information would include (among other things) B's assertion that his clock reads 12 sec. after time zero, and TR's report that he has received only 6 of the 12 pulses emitted by A during the time of the trip.

In order to emphasize the absolute and non-negotiable nature of the measured distance between A and B, we will now introduce a measuring device in the form of a wheel with a circumference of one meter. The wheel rotates on an axle that is fixed to TR's lab and it also rolls in a track along the A to B route. A device on the wheel's periphery makes a mark at every meter of the entire AB track — 1.8×10^9 meter marks along the 6 light second track. The same marking device allows TR's lab to record 1.8×10^9 wheel rotations during the AB trip.

The purpose of this wheel scenario is to show that TR and AB observers should be in agreement about the AB distance because they used the same measuring device at the same time to arrive at a single distance of 1.8×10^9 meters. Of course, relativists might try to claim that TR perceives the dimensions of the wheel in a different way. This, however, might be difficult to sell, since it would imply that one wheel in one location at one instant in time must have two different and equally valid sets of dimensions. Incidentally, it is understood that some of this model's assumed hardware, like the AB track material and the wheel structure, is mechanically impractical considering the distances and speeds that are used. Just presume that the wheel is composed of a virtual reality material that would not explode when rolling along at a speed of 1/2 c.

It should be established at this time that all value determinations concerning distance are properly derived only at the source in their reference frame of origin. So, numbers like the A to B distance, distance between pulses from A and B, and wave lengths of light originating in the AB system are all unchangeable or absolute values that should not logically be vulnerable to alteration by observers in other inertial systems. TR, as a moveable recipient, can conceivably claim that he perceives time and its related values, like frequency and velocity, in a way that is different from the perceptions of AB observers, however distance values from the AB system are not directly perceivable to TR. He can only accept the distance values furnished by the AB observers, or re-calculate the numbers using his already distorted perceptions of time as a basis for those calculations. In other words, measured distances in one system are not negotiable quantities that can have different values for different observers.

The next model addition will be the introduction of 6 more physics lab substations (labeled S1 through S6) that are evenly spaced and located at one light second intervals along the AB track. Please become familiar with the diagram on page 51 and note that there are a series of 3 simultaneous events occurring at each of the 7 labs located at A, B, and the substations. Each triple event represents a confluence in which

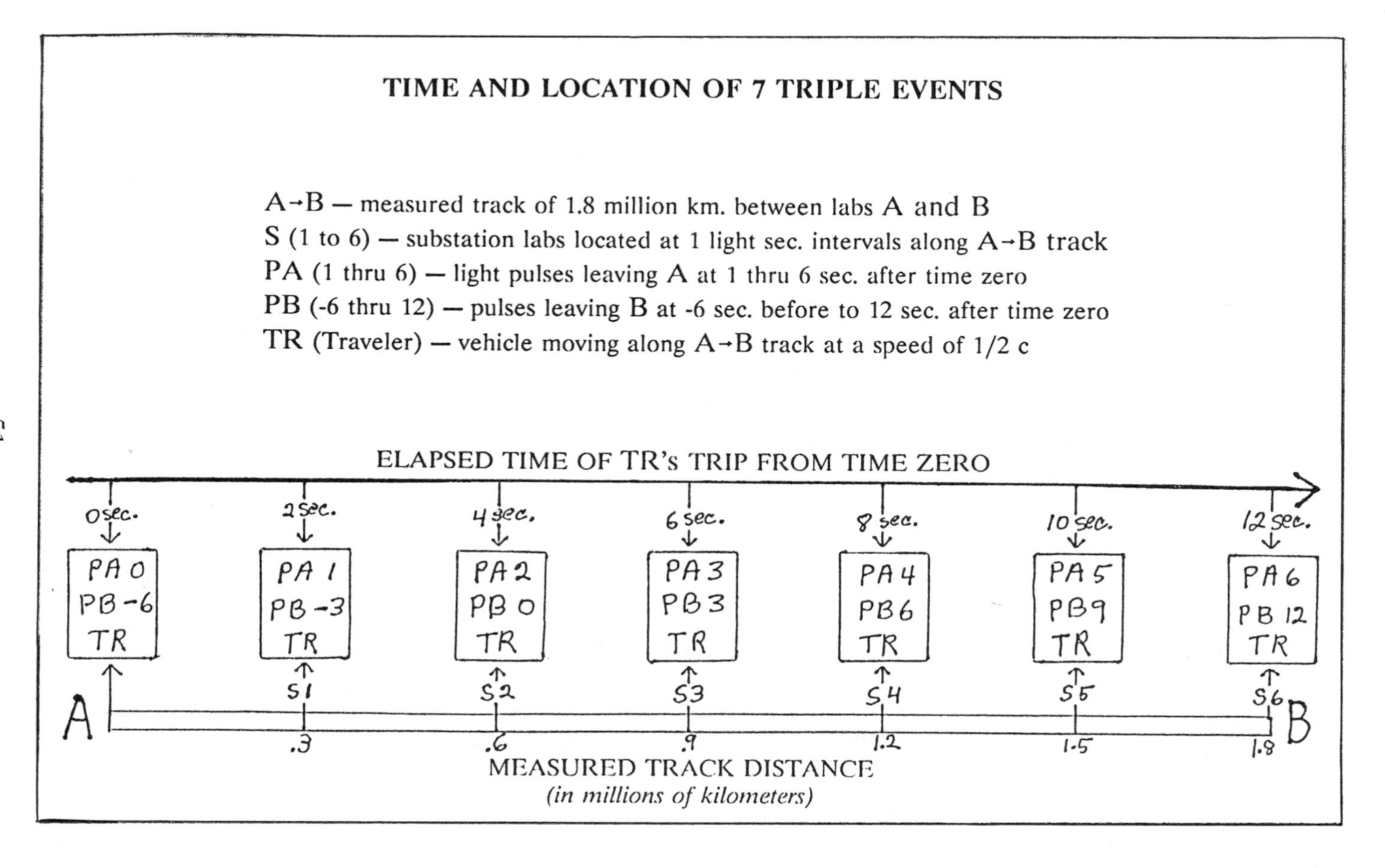
TIME AND LOCATION OF 7 TRIPLE EVENTS
A→B — measured track of 1.8 million km. between labs A and B
S (1 to 6) — substation labs located at 1 light sec. intervals along A→B track
PA (1 thru 6) — light pulses leaving A at 1 thru 6 sec. after time zero
PB (-6 thru 12) — pulses leaving B at -6 sec. before to 12 sec. after time zero
TR (Traveler) — vehicle moving along A→B track at a speed of 1/2 c
ELAPSED TIME OF TR's TRIP FROM TIME ZERO
0 sec.
2 sec.
4 sec.
6 sec.
8 sec.
10 sec.
12 sec.
PA 0
PB -6
TR
PA 1
PB -3
TR
PA 2
PB 0
TR
PA 3
PB 3
TR
PA 4
PB 6
TR
PA 5
PB 9
TR
PA 6
PB 12
TR
S1
S2
S3
S4
S5
S6
A
B
.3
.6
.9
1.2
1.5
1.8
MEASURED TRACK DISTANCE
(in millions of kilometers)

TR and light pulses from A and B come together at a specific lab location. For example, at 4 sec. into the experiment (4 sec. from time zero), the number 2 pulse from A, the number zero pulse from B, and TR, all arrive simultaneously at the S2 lab. The importance of the event combinations lies in the fact that these undeniable realities are actually experienced at the same instant by TR and the observers along the AB track, so they must necessarily be perceived in the same way by all observers, including TR. Working from this platform of agreement, the manipulation of logic imposed by the constant light speed postulate can be more clearly shown by models involving paired comparison of closing velocities. However, to avoid lengthy side trips, the pursuit of details in this argument will be left to the discretion of the reader.

Now, getting back to the perceptual consequences of relative motion between two frames of reference as interpreted by various observers, we can trace the events that occur during TR's constant 1/2 c trip from A to B. Recall that both A and B send out light pulses at 1 sec. intervals while TR traverses the 6 light sec. between A and B.

During the single time interval of the AB trip (12 sec. as determined by AB observers), TR receives 18 light pulses from the B direction but only 6 from the A direction according to the following explanation; A sends out 12 pulses during the 12 sec. it takes TR to travel from A to B. Six of these pulses pass through TR and the other 6 are in transit between A and B when TR reaches B at the 12 sec. mark. In the other direction, TR impacts the 6 pulses that are in transit from B to A at the moment the model begins (time zero), plus the 12 additional pulses that are emitted by B during the 12 sec. of TR's A to B trip. Thus, a total of 18 B pulses are impacted by TR. In terms of distance along the AB track, TR impacts A pulses every 300,000 km and B pulses every 100,000 km. This means that TR and the light pulses from B approach each other with a closing velocity that is 3 times as great as the closing velocity between TR and the light pulses from A. Recall at this point that the distance between successive 1 sec. pulses from A and B is a constant 300,000 km. Now, the special relativity postulate holds that TR must perceive the closing velocity of all light pulses from any direction as being 300,000 km/sec. (c). To resolve this impending conflict, TR first considers the 6 pulses from A in the following manner. He sees that it will be necessary to change the 12 sec. duration of the trip to a lesser number by changing the function of his own clock. If he did not make this adjustment, he would be in the position of perceiving 6 evenly spaced A light pulses (300,000 km apart) in a 12 sec. period, meaning that he would be receiving light at the unacceptable speed of 1/2 c. So, by slowing his own clock and reducing the total trip time, TR is able to raise the unacceptable 1/2 c closing speed toward the proper speed of c.

Rather than being concerned with the math equations and the use of the L.C. factor in these changes, it is more important in the current argument to remember the direction of these time changes that TR makes because of the problems created by the nature of the pulses from A and B.

Now, if TR's only contact with the world outside his frame of reference was the 18 pulses from B rather than the 6 from A, things would be different. His observations would then indicate 18 light pulses (with 300,000 km spacing) arriving in a 12 sec. period with an unacceptably high closing velocity of 1.5 c. So, following the previous reasoning, TR would have to speed up his own clock and lengthen the trip duration in order to reduce the closing velocity with the B pulses from 1.5 c toward the acceptable value of c. However, since this model was arbitrarily set up to include pulses from both A and B, TR is confronted with the problem of having two sets of equally valid trip duration numbers (one higher and one lower than the determination of AB observers). He cannot logically report two values for the duration of a single trip without revealing that his methods are in error. Of course, this paradox or impossible duplicity is the result of the off base assumption that two observers receiving light from a single source must receive that light at the same speed of c regardless of relative motion between the two observers.

It should be pointed out that in this model TR is a "naked observer" in that all of his instruments are on the outside of the vehicle. In other words, there is no volume of light transmitting space medium surrounding and traveling with TR and his instruments. This point is emphasized for the purpose of avoiding the same type of misinterpretation that came out of the Michelson-Morley experiments when the static nature of the medium in the vicinity of this earth bound experiment was assumed to be an absence of light transmitting medium.

Nearing the conclusion of this A-B-TR discussion, it seems appropriate to summarize the non-relativistic reality of the model in the following way; the one sec. interval light pulses go back and forth between A and B with a speed of c because of the fact that the light transmitting medium between the two is at rest relative to both A and B. Since TR is moving relative to this medium, he receives the 6 pulses from A at 2 sec. intervals and at a perceived and actual speed of 1/2 c. TR also receives the 18 pulses from B at 2/3 sec. intervals and at a light speed of 1.5 c. Both of these sets of recordings indicate a trip duration of 12 sec., that of course matches reality as well as the clock readings of all observers, including TR's. TR is able to perceive these light velocities above and below c because he is a "naked observer" in that he is not surrounded by a cocoon of light transmitting medium that moves along with him.

Furthermore, this thought experiment could have been constructed so that TR was accelerating during the A to B trip, in which case the math would have been a little more complicated, but all of the important paradoxes, denials, affirmations, and conclusions would have been exactly the same. The relativists' idea that accelerated motion should cause anything more than the expected algebraic changes in the results of the single velocity model is probably best interpreted as an attempt to theoretically bind gravitation to light speed perception by using the equivalence of gravity and acceleration to justify the claim that acceleration has more than the normal mathematical effect on light speed perception. More will be said about this in a subsequent chapter.

The essence of the preceding model involves the construction of a series of paired events in which the postulates of special relativity demand that there should be a certain symmetry. To the extent that asymmetry has been demonstrated, it can be seen that these postulates are in error. To summarize it in another way; we have demonstrated that two observers in relative motion, receiving light from a single source, are not interchangeable as far as both perceiving light at the same speed is concerned. This is because the speed of light is tied to the movement of its transmitting medium which cannot have a single velocity relation to both observers.

In the non-scientific literature, presentations of special relativity are frequently adorned with such laudatory phrases as "simple but beautiful" and "having more depth and vision than first reading would reveal." This type of description seems to reflect an abundance of sentimental popular acceptance and possibly less than the normal standard of critical skepticism.

At any rate, trying to trap an adroit relativist in an untenable paradox is somewhat like a combined semantic and mathematical chess game. Whenever the defensive player seems to be cornered, he escapes by arbitrarily changing the squares on the board, the shape of the players, or even the rules of the game. Imaginative thinking emerges as the most important factor in winning. These are the frustrations of trying to pin down such a versatile and elusive adversary as a committed relativist.

When the scientific community realizes the interpretive errors that were made one hundred years ago following the Michelson-Morley experiments, it will finally be recognized that the speed of light is media determined, and that this medium is capable of differential motion among its various volumes. Then it will become obvious that the perceived speed of light by any observer must be a variable value determined by that observer's motion relative to those volumes. Special relativity would then fade away, and the contrived mechanisms of time dilation, variable ageing, and length contraction would all fall into place as unreal and unnecessary accommodative assumptions.

The idea that c must represent the upper limit of all speeds of mass through space seemed to grow out of the special relativity predictions concerning time, length contraction and mass in the following way. The theory holds that mass increases and length contracts as velocity nears c in such a way that mass would become infinitely large when velocity reached c. Since infinite mass is an illogical concept, then it must be impossible for mass through space to attain a velocity of c. Thus, c became an important universal constant that might logically be suspected of harboring even more undiscovered applications. Of course, Einstein subsequently applied the light velocity constant to the behavior of gravity in the general theory of relativity. Historically, and even today, there has been far too much mystical importance attached to c. Actually, it is only the speed of light through space medium. Even in this context it is not a constant, but rather, a number with a small variability range to match the small variability range of the texture or density of the space medium that determines its value. So, the eventual negation of the time, length, and mass alterations of special relativity will remove the reason for imposing c as a speed limit for mass.

There have been a number of observations in the last few years suggesting that mass is in motion at a speed above c (super luminal velocity). These observations have usually concerned quasars and active galaxies, but at least one has been within our galaxy. The number of such observations is bound to increase because the reality of the situation is that there is no limit to the velocity of mass through space, which is another way of saying that there is no limit to the relative velocity between two contiguous volumes of the universal entity. Of course, some of the super luminal observations will prove to be aberrations, but our ever increasing range and clarity of detection will soon overwhelm those relativists who try to rationalize away each new discovery of super luminal velocity.

To me, it seems likely that some of the best and most easily verifiable examples of super luminal velocity should be found in our own galaxy among the class of rotating objects called eclipsing binaries — those composed of one or two very dense objects in a tight rotation with a very short rotation period of hours or days. Super nova like events might also be a promising area in which to search for these velocities. Hopefully, this parade of super luminal velocity observations will assist in the early relocation of c to its logical and proper status of being only the average velocity of light in a vacuum.

Before leaving the subject, I will comment on a couple of experimental situations that are frequently cited as evidence in support of special relativity.

In 1971, two atomic clocks were placed aboard two east and west bound, around the world jet flights. The east bound clock lost 59 nano seconds while the west bound clock gained 273 nano seconds (1 nano sec. = 10^{-9} sec.). This is considered by some as evidence for the time dilation aspect of special relativity. My first reaction to these clock aberrations is that they are most likely due to altitude gradient variations in the cohesion and density (or concentration) effects of gravity. And it makes no difference whether the clock is mechanical or is related to radioactive decay time as in the case of atomic clocks. This altitude gradient of gravitational effect would pertain in either case, especially since such tiny temporal aberrations are involved. The other observation to be made is that special relativity predicts that both the east and the west bound jets should age less, or lose time, relative to the stay at home clock. So, the observed nano second clock aberrations are not consistent with this prediction.

The other relativity supporting situation involves a muon decay experiment. A quotation from the Scientific American publication, Relativity and Its Roots, is as follows; "A striking confirmation of the special theory of relativity was found by observing electron like particles called mu mesons, or muons, that are created in the atmosphere by cosmic rays." The essence of the situation concerns an unexplained lengthening of the predicted decay time of fast moving muons as they approach earth. This was interpreted as being due to special relativity type of age or time retardation.

First, the muon decay model is based on some shaky qualitative and quantitative assumptions about the nature of cosmic radiation. The time alteration route for explaining this observation involves using relativity predictions in a questionable clock — observer — velocity scenario. The several conclusions thus derived also seem to employ a type of circular argument in which one part of a linked package of relativity predictions is used to establish the validity of another part. This, however, is more or less standard logical procedure that is found in most special relativity models involving rockets, traveling twins, or whatever.

Finally, I would suggest (as in the jet plane experiment) that the altitude gradient of gravitational cohesion and density would be a more likely explanation for variation in muon decay time than relativity related time alteration. At best the "striking confirmation" statement would be a commendable, but weakly based and improbable, effort to support the theoretical work of a respected and honored scientist such as Einstein. At worst, it would represent an unwarranted rationalization and exaggeration of a low probability situation into a striking confirmation.

I am not aware of any positive observations or experimental evidence that can currently be cited as favoring the non-relativistic behavior of light and time; however, such evidence could eventually emerge, possibly as the result of satellite experiments involving the parameters of light as outlined at the end of the light transmission chapter.

There are a few philosophical comments to be made at the conclusion of this chapter, beginning with a statement of reality as follows.

In the universe, there exists at this instant an infinite number of event locations, each having an infinite number of spatial coordinates relative to the other events. This statement reflects the ability of the human brain to have <u>instantaneous</u> <u>deductive</u> <u>understanding</u> that covers the entire spatial expanse of the universe with no message delivery time lag.

In a brief digression, I will try to describe the concept of time as a preliminary to subsequent statements. Functionally, the universe might be characterized as a patchwork of events, within each of which there is a large number of secondary events. The combined activity (probably both sequential and directional) of these secondary events causes a single increment of activity or change in the parent or containing primary event. This establishes the hierarchy nature of activity in the universe – multiple levels of events within events.

Time is a man made concept or tool developed for the purpose of recording and recalling the relationships among all the events on a particular hierarchy level, as well as numerically relating the members of different levels. Thus, time is not an entity like the event systems, but only a devised recording system that is by definition, incapable of natural variations, as is the case with all the variable parameters among the event systems it attempts to describe and relativise. Distance, on the other hand, is a value designation used to describe the spectrum of human perceptions caused by the variable distribution patterns resulting from mixing of space and mass phases of the universal entity. Therefore, time and distance are two different classes of man made measuring tools. They are functionally similar only in the fact that time relates to events the way that distance relates to the universal entity.

This brief commentary on time is not meant to be comprehensive or anything more than marginally close to reality. It is included at this point only for the purpose of lending some credibility to the opening presumption in the next paragraph.

If we presume that time passes the same way for everyone, then having spatial and motion knowledge of two past events allows the extrapolation of that information into an understanding of their present related status without recourse to current perceptions and their variable message delivery time aspects.

For some reason, earlier in this century, there developed a heavy reliance on perception as the criterion for reality. Perception usually involves message delivery time between observers and events, which becomes important in determinations concerning the media dependent speed of light and/or variable motion between volumes of media. If there is also motion or accelerated motion between observers, then defining a single reality that is common to all observers becomes mathematically involved but not impossible. It is only when errors in interpreting time lapse messages due to faulty assumptions about light speed perception enter the picture that it becomes necessary to have more than one set of figures to describe the reality of a single event.

The existence of simultaneity among all observers and events, regardless of their relative motion, is a reality that has been clouded by elevating current perception to a higher level of reality determination than the process of deductive understanding. Even this, however, would not have resulted in today's double standard of event description if the light speed errors had not found their way into the perception model.

The history of human reasoning concerning methods of deductive thinking and sequential logic seems to be marked by fashionable trends that fluctuate over the decades like clothing styles. Somewhere around the turn of the century the idea that there was no limit to the power of the human intellect seemed to flourish and gain in popularity. It is probably quite natural to crave the feeling of relaxed security that can flow from the delusion of intellectual dominance over everything in one's environment. Possibly the exaggerated importance of human perception grew out of such a comfortable mind set in which nature was fantasized as being malleable and responsive to the moods and desires of the human intellect. Recall the saying — "beauty is in the eye of the beholder." The idea that such basic concepts as time, distance, and mass could logically be varied or have their values manipulated by the force of human mentality may have subconsciously energized the type of thinking that has allowed the excesses of relativistic theory to distort our concepts of the physical world and its behavior.

EQUILIBRIUM AND HIERARCHY ASPECTS OF AN AGELESS UNIVERSE

I realize that this chapter may seem somewhat out of place in a section of the manuscript that deals more with basic physics such as the dissection of phenomena like light and forces, and the small unit structure of mass and the space entity. The reader might consider this change of pace as a brief return visit to the cosmos. It is being included at this time partly for lack of a better location, and partly because the subject might be more easily understood and accepted considering the relevance of some of the preceding material concerning the space phase of the universal entity.

Possibly the single most impressive overall observation of today's universe is one of dynamic energy production or exchange on a very large scale. Luminosity in the visible and other wave length ranges suggests that most of this energy production is occurring within objects that are identifiable as stars.

Probing into this observation a little deeper reveals that the micro mechanics of this activity is almost entirely of the atomic fusion type rather than chemical. Also, that every step of the fusion process is accompanied by a depletion or disappearance of mass that is equivalent to the energy being produced and radiated away from that location. Less obvious, but equally true, is the fact that this fusion activity, acting in concert with gravity, causes constant change in the gross appearance (shape, size, density, luminosity, etc.) of the larger organizations of universe mass like stars and galaxies. If all of these gross and micro changes were to continue indefinitely, then the appearance of the universe at some distant future time could not resemble today's appearance. The fusion caused shift in the universe chemical composition and the depletion of mass would eventually result in a cool, dark, inactive universe with no semblance of today's picture.

Let's presume that it has been established that the universe is infinitely old. The logical definition of infinite age requires that the object in question remains the same in appearance and composition etc. throughout all time. Clearly, the fusion burnout described above would not be consistent with an infinitely old universe. By the same reasoning, today's dynamic appearing universe with its progressive one way destruction of mass, could only have begun a short time ago rather than in the infinite past.

The only way out of this dilemma is to assume that all of today's apparently uni-directional gross and micro changes are actually cyclic reactions in which the end products are converted into the raw materials for these ongoing reactions. In other words, the cyclic refueling of all the gross and micro changes that we currently observe is the only way to perpetually maintain the necessary equilibrium which is a defining characteristic of an infinitely old universe. Admittedly, our current lack of observational evidence in this area will make the selling of this circumstantial scenario somewhat more difficult.

Let's first analyze the implications of this refueling idea as they affect things on the micro or fusion level of activity. There will be some unavoidable repetition of my 1993 writing at this point.

In trying to derive numbers for the chemical composition of the universe, we can start with solar composition estimates of 75% hydrogen, 24% helium, and 1% heavy elements. If the sun is representative of main sequence stars, which account for the bulk of fusion activity in our galaxy, then these figures would hold for the universe to the extent that galaxies resemble each other. Keep in mind, however, that I am not trying to characterize all mass in this manner, but only that portion of universal mass that is actively involved in the hydrogen to helium to heavy element fusion process, which is the portion we identify as luminous stars. It is estimated that the solar hydrogen to helium fusion rate is 6×10^8 tons per sec., and that 1% of this quantity goes into energy. In other words, 1% of the involved mass disappears. Considered on a universe scale, it is obvious that this directional reaction would cause the gradual depletion of hydrogen and the accumulation of heavy elements until the 75-24-1 ratio was reversed to the point that fusion no longer occurred. Starlight would cease and the universe would become a dark inactive entity. It would be possible to construct a time frame for this eventuality in billions of years by using the fusion rate, the number of stars in the universe and the estimated mass of the universe. However, this would lead to considerable quantitative guessing, which is not really necessary as long as we assume that the universe is infinitely old. It becomes obvious that maintaining this fusion reaction with its 75-24-1 ratio features for a very long or an infinitely long time would require both a dissolution of heavy elements and a reformation of hydrogen in order to complete the cycle and thus permanently ensure equilibrium.

Incidentally, as an interjection at this time, it should be noted that observing distant galaxies is the same as looking back in time. So, to the extent that stars in a distant galaxy seem to have the same luminosity features as the stars in our galaxy, they are probably also similar in chemical composition and fusion activity. This would suggest that there has been very little, if any, change in stellar chemical equilibrium over a period of several billion years. This, in turn, would give strength to the argument that stellar fusion is a closed and self sustaining type of cyclic reaction. Of course, the standard Big Bang theorists would prefer to ignore any such suggestion of long term universal chemical stability. To them, this 75-24-1 ratio must logically be only a transient picture in the one time expansion, development, and final disposition of a limited universe.

At this point I will devote some speculative discussion to the dissolution of matter — the natural conversion of mass from the atomic to the pre-hydrogen status.

The top atomic numbers that are normally achieved in main sequence stellar fusion ovens are elemental size limits about which we have no hard or observable knowledge. For the moment, let's presume that the heaviest element formed in stellar fusion is somewhere between carbon and iron. Then, it follows that there must be a mechanism, possibly periodic episodes, that changes some of these medium range elements back to a phase that is not recognizable as mass, but at the same time is a precursor to protons or hydrogen. This could logically be called the space phase of the universal entity.

It is generally understood and accepted that the disappearance of mass at every step of the stellar fusion chain of reactions is accompanied by the instantaneous appearance of an equivalent amount ($E=mc^2$) of radiative energy. The standard conclusion is that mass was transformed into energy. I would suggest the possibility, however, that mass changes to a non-mass space entity form which may be both larger in volume and smaller in potential energy (on a per unit basis) than the previous atomic state. This volumetric change in mass units would introduce an expansive type of disturbance into the medium surrounding the event — a disturbance that we interpret as light or some other wave length of radiant energy. So, rather than being the embodiment of the disappearing mass, energy would then be seen as a disturbance produced as a byproduct of the alteration or downgrading of the internal motion parameters (possibly rotational types of motion) of a mass unit to a space entity unit. Other areas of this writing contain more speculative details on the micro mechanics of this conversion of mass to the space phase of the universal entity.

Of course, there is no way to accurately deduce either the mechanism of this elemental dissolution or its subsequent reconversion into hydrogen. However, speculating on the dynamics of the situation, if the units of the space entity are characterized by an inter unit attraction whose strength is determined by the parameters of the curved motion patterns within each unit, then this universal attractive force (UAF) would act very gradually to cause a reduction of volume occupancy or condensation of contiguous units.

To emphasize the very gradual but universe wide nature of space entity densification, consider the following; as part of his steady state theory some decades ago, Fred Hoyle determined that the required rate of hydrogen replacement of stellar depletion would only entail the creation of one hydrogen atom per liter of universe volume every 10^{12} years or so.

The preceding paragraphs have mainly been concerned with trying to picture the micro mechanisms involved in the production of hydrogen and the dissolution of heavy elements. The next step in the discussion should logically concern the gross or macroscopic aspects of these reactions and their time frames. For example, the location, size, and time interval features of the relatively rapid expansive (or explosive) types of events that result in heavy element dissolution should be looked into. However, before getting into these large scale implications of equilibrium, I wish to spend some discussion time trying to develop a better baseline concept of the natural organization of everything.

Consider the following as a logical probability. Operating in an infinite volume, the UAF causes the natural organization of the universal entity into systems within systems. If the universe is a fractal type of hierarchy system in which there are multiple levels of organization, we can then think of stars and galaxies as being members of two contiguous levels in that hierarchy. Stellar systems would be members of the intra galactic level of organization, and galaxies would be members of the next larger or intergalactic level of organization.

First, examining the smaller intra galactic level of organization, I would suggest that the 100 billion (10^{11}) or so stars in our galaxy represent only a tiny fraction of the total number of contracting, rotating, flattened spheres that <u>totally</u> occupy the galactic volume. The total galactic population of these rotating units, including visibly luminous stars, is maybe on the order of 10^{13} or 10^{14}. These non stellar units would have a wide range or gradient of decreasing density and increasing volume when compared to typical stellar volume and density. So, within our galaxy, there are a large number of these proto-stellar units containing stellar mass and pre-mass quantities that are completely invisible to us because they have not yet reached the threshold of density and temperature that would trigger hydrogen fusion and luminosity.

The composition of this myriad of pre-stellar, semi spherical units would probably be mostly hydrogen and pre-hydrogen space entity units with a sprinkling of other mass in all phases from gas to dust to particles to planetary sized objects, but no items large enough to support fusion. The total quantity of mass plus pre-mass substance in one of these non-stellar units would probably be comparable to the mass of a large star, since these units eventually gravitate themselves into stars.

Turning now to the next larger or intergalactic level of organization, I will make the somewhat unorthodox suggestion that quasars or the nuclei of "active" galaxies are the objects that are comparable to stars on the lower level. Spiral and elliptical galaxies are fuzzy, poorly demarcated, less dense conglomerate structures that are in the process of slowly contracting to quasar status. Galaxies, then, are comparable to the previously described invisible pre-stellar rotating structures at the intra galactic hierarchy level. The fact that galaxies are the predominant objects that are visually identifiable at this larger level of organization leads to the common misconception that galaxies and stars are comparable items on these two contiguous levels.

To further explain or emphasize in a slightly different perspective; a quasar is preceded by a much larger rotating unit composed of gas, dust, and billions of particles (stars) that are destined to gravitationally compress into a quasar. Similarly, a star is preceded by a larger rotating unit of gas, dust and particles that will become a star by gravitational compression. However, since none of the objects in this pre-stellar organization is large enough to support hydrogen to helium fusion, the unit remains invisible to us. Conversely, in the comparable situation at the larger hierarchy level, billions of the particles (stars) in the pre-quasar rotating unit are engaged in fusion, thus making the whole comparable unit (or galaxy) impressively visible to us.

The preceding comments on comparative anatomy gives us a picture of two different levels of organization that match each other fairly closely concerning their qualitative structural features as well as in the matter of volume to structure ratios. It would also be logical to expect similarity in behavior and events on both levels with an event or organizing <u>time frame</u> that would vary in the same way as the sizes of the structures vary.

For example, quasars should eventually undergo some kind of expansion comparable to the several types of explosive or expansive events that will happen to all main sequence stars. Considering that a super nova occurs in our galaxy about every 50 years, we might expect a similar quasar event to occur about every 5 million years (within a radius of 100 billion light years or so). Thus, the obvious rarity of these events would make their lack of human detection understandable. Similarly, the slowness of any current quasar expansion would make its earthly detection difficult.

The duration, however, of a quasar expansion might not be as many times longer than a super nova event as comparison of their respective volume settings would indicate. If the quasar's much greater temperature and density compression features resulted in the dissolution of heavy elements wherein the products of the reaction (energy, lighter elements, and space entity units) had a volume occupancy that was some 5, 10, or 100 times greater than the volume of the original heavy elements, then the ensuing sequence of temperature and pressure increase, chain reaction, and explosion could be magnitudes more violent and more rapid than a comparable super nova explosion. Recall that we have previously disposed of the rule or law that forbids super luminal velocity in such explosions. So, if the expansion rate in a typical quasar expansion was 100 times greater than the speed of a super nova explosion, then the duration of a quasar explosion might be 100 rather than 10,000 times greater than the several months duration of the active phase of a super nova event.

There is some precedent for presuming that larger objects might undergo different and more violent expansive reactions than smaller ones. Looking back at stellar expansions, we see that increasing the object size by a factor of 3 or 4 changes the type of expansion from a slow symmetrical red giant formation to a much more violent and rapid explosive reaction called a super nova.

The mechanism of an explosion, as suggested in the previous discussion, deserves a little more elaboration at this point. Whenever the increase in temperature and density compression features cause the threshold of an atomic reaction to be passed, the stage is set for a possible chain reaction. However, if the volume occupancy of the end products of the reaction is similar to or smaller than that of the original elements, then there will only be a steady and prolonged radiant energy production that parallels the gradually increasing temperature-compression features driven by the UAF. The usual long term fusion luminosity of main sequence stars would be the typical example of this non-explosive type of reaction. If, on the other hand, the volume occupancy of the reaction products is greater than that of the initial elements, then the sudden volume change would further increase temperature and pressure which would in turn speed up the original reaction. A rapidly increasing, self-feeding chain reaction would be established causing an expansion which would outwardly propel everything peripheral to the central reaction. This volume occupancy scenario is probably applicable to all chemical as well as nuclear explosive events.

As an incidental speculation, quasars are probably conducting their heavy element fusion activity at temperature and density values that are far above those that would cause violent explosion in an isolated stellar setting. Obviously, the quasars' much greater self compressing gravitational ability permits the containment of this particular range of reaction parameters, just like the sun is able to gravitationally contain its particular type of hydrogen fusion explosion within its core with no resultant distortion in the overall shape or size of the sun.

Similarly, the thresholds of quasar expansion must involve qualitatively different atomic reactions as well as explosion inducing parameters of those reactions that are vastly different from and greater than those at the stellar level.

Without going into justifying arguments at this time, it should be noted that much of the preceding discussion assumes that the theoretical "black hole" type of gravitational collapse is a non-reality. Please see the sub atomic chapter and the epilogue for more on the misconceptions that allowed this erroneous black hole model to be popularized.

There are two things that make the concept of quasar expansion both logical and probable. One is that the structural similarity between contiguous hierarchy levels suggests that there should also be behavioral similarities. The other reason is that the idea of an infinitely old universe demands (as a defining characteristic) that there be a perpetual recycling of structures at all hierarchy levels in order to maintain the necessary gross structural equilibrium.

Now, getting back to the subject of the formation and dissolution of heavy elements. It is generally assumed that the formation of all elements from helium up to uranium occurs in stellar events. Within this limited framework of theoretical mass creation, it is usually presumed that the lighter elements, up to carbon or so, form in the normal main sequence stellar ovens, and that the heavier ones somehow occur in stellar expansive events like super novae and red giant formations.

I would suggest that the higher temperature and greater density of quasar like structures would be a more logical event location for the fusion formation of the heavier elements. Without trying to guess the atomic number type of dividing line between the processes, I would only presume that there is probably considerable overlapping in the quasar and stellar systems of elemental creation.

Further, I would suggest that all the explosive and expansive events among stars and quasars play no significant part in the fusion building of heavier elements. In contrast, these expansive events with their tremendous energy radiations would more logically fit into the arm of the equilibrium cycle where all the heavier elements are disassembled, downgraded, or structurally transformed into items that have less potential energy and occupy more volume on a unit for unit basis than their preceding or parent elements. Thus, expansive events at all size levels would be the main avenue for completing the equilibrium cycle by disassembling mass and reconstituting the space entity pool, or pre-hydrogen sink, that accounts for the entirety of the universe volume that is not humanly perceived as mass.

In trying to describe the hierarchy nature of structure and events, I may have conveyed the erroneous impression that there are sharp lines of distinction between the quantitative features of structures and events on different levels. Of course, the reality is that there is a great deal of curved gradient type of blending and overlapping of all the temporal, volumetric, and mass-structural values involved in contiguous hierarchy levels of interaction. Nature seems to prefer curvature, gradients, and variables in all values and relations; whereas humans seem to be constantly driven to impose constants, limits, and straight line relations despite the clues of curvature that nature displays. Picturing two side by side sine curves overlapping each other in the 2-3 standard deviation range may help in visualizing the way these structural and event values probably tend to blend between contiguous hierarchy levels.

One example of overlapping structural organization on different hierarchy levels is seen in the form of "globular clusters" of stars in our galaxy. These clusters contain thousands of stars and clearly represent intermediate gravitationally contracting units that are many times smaller than galaxies and many times larger than typical single or binary stellar units of organization. Conceivably, there could even be some kind of expansive event that periodically occurs in these clusters. The variables concerning velocity, periodicity, violence, and duration of such an expansive event would obviously lie somewhere between those of super novae and quasar expansions.

In the following model and review, I will try to show that our knowledge of the earth's composition and origin can be used to reinforce the previous arguments favoring the hierarchy nature of structure and event organization.

As it happens, most of the earth is composed of iron and a variety of medium and high range atomic number elements. If one accepts the previously discussed theory that these heavier elements are more logically created in the larger (quasar like) hierarchy level ovens than in the smaller stellar ovens, then some interesting implications concerning universe age and recycling activity will unfold as we trace the origin of earth's heavy element composition.

First, by way of a baseline review, it should be noted that the most recent historical event in the history of the solar system was probably some kind of medium or slow speed expansion of the sun's stellar predecessor that was less violent than a super nova. This expansion spread the parent star's mass into a flattened sphere whose diameter was close to that of today's solar system. Natural rotation flattened the sphere into a disc. The diversity of elements then caused the disc to segregate itself into concentric rings of gas, dust, and particles based on the distribution of atomic numbers or atomic weight. This was followed by the gravity driven globular concentration of the mass in these various rings into the rotating disc like system of planets, moons and asteroids that characterize today's solar system. This fairly standard picture of solar system formation is reinforced by the observation that all planets are revolving around the sun in the same direction in addition to occupying the same thin disc shaped plane.

The most recent stellar expansion that created the solar system may have been preceded by several generations of stellar expansions, reorganizations, and contractions. Much of our solar system's lighter element content was probably produced by fusion in the stellar ovens of these earlier generations of stars, but earth's heavier elements were just carried along and redistributed by these stellar events. Earth's heavy elements were already present in the relatively large pre-stellar invisible pool of rotating gas, dust and particles that condensed to form the first star in the above mentioned series of stellar generations.

Meanwhile, our galaxy (which contains this solar system drama) is in the process of gravitationally condensing in a spiral rotating fashion. The solar system mass is located in the outer portion of this spiral galactic structure that is in the mid range of a very prolonged rotating condensation or transformation from a much larger proto-galactic volume status to an eventual quasar like object. So, earth's heavy elements, having never been part of our galaxy's core, must have formed in the quasar like core of a previous galactic generation. This older quasar expanded and distributed its wide range of fusion created heavy and light elements to the proto galactic volume that has recondensed into today's spiral galaxy picture. Therefore, earth's composition suggests that some of its elements were created in stellar ovens some 10^9 or 10^{10} years ago, while others must have originated in quasar ovens possibly 10^{11} or 10^{12} years ago.

The purpose of the preceding rambling discussion is to

establish the probability of cyclic recurring generations of galaxies, similar to the recognized and accepted cycles of stellar generations. Of course, this would logically push the minimum age of the universe into the trillions of years — 10^{12}, 10^{14} or very possibly an infinite number of years.

It might be possible to characterize the cyclic behavior at both stellar and galactic levels in a general way as follows; variable sized mass containing volumes tend to undergo repeated cycles of contraction, expansion and re-contraction, complete with reorganizational sharing of the expansion phase mass between contiguous mass containing regions, thus allowing perpetual random variety in the geography of density foci that precede the next cycle of contraction.

This chapter, as the reader may have noticed, is somewhat digressive and fragmented with considerable over explaining. My excuse for this rambling is that the interwoven nature of the array of events has made the organization and writing into a difficult task — like trying to describe a machine with too many moving parts.

SUBATOMIC STRUCTURE AND BEHAVIOR

Completing this series of critical arguments calls for some discussion of the conceptual frailties that are found in today's widely accepted view of atomic structure. The subsequent material contains abundant speculation but very few facts. So, by skipping this chapter, the reader will not run the risk of missing a bold new detailed picture of the atom's interior. Considering the make-up of the universe at the smallest invisible level involves even more probability guessing than large scale universe modeling. My ideas on alternative ways to conceptualize structure and behavior at this level are contained in a group of generalized speculations and arguments that have no accompanying observations and few if any details or specifics.

Starting this chapter with such a broad disclaimer, one might logically ask; why, then, do you feel the need to throw rocks at the standard atomic model if you have no coherent theory to take its place? My answer would be that pointing out problems with the current model should not necessarily carry the obligation to offer a more realistic replacement. Let's make intellectual room for honest answers like "I don't yet know." If it turns out that the current model of intra atomic structure and behavior is destined to collapse, then this temporary void in knowledge and understanding should be gracefully accepted by everyone. The implication is that theorists should resist the temptation and pressure to prematurely replace one flawed concept with another.

Most of us, including scientists, instinctively agree that there must be some kind of structural organization and pattern of behavior that is occurring within that very small atomic volume. Flashing a picture of the standard atom (nuclear protons and neutrons surrounded by shells of electrons) on the screen inside our heads is both useful and potentially misleading. There seems to be a natural human tendency to gradually assign too much reality identification to these man made concepts, especially when they have been around for something like 100 years or so. Repetition has a tendency to harden or institutionalize this subatomic structural nomenclature into items that are popularly viewed as established real particles rather than useful conceptual models as they were originally conceived. Accepting the standard descriptions of neutrons, protons, electrons, and their interrelated behavior would be the least antagonistic course for me to follow, but it would not be completely honest. Also, I realize that my skepticism about the standard atom will do nothing to dispel the reader's suspicion that I am simply anti-everything. However, my basic idea that the universe is composed of a single universal entity having a very wide range or gradient of perceptual phases with no intervening empty spaces is clearly not consistent with discrete units of mass in a matrix of empty space. The standard concept demands that there must be sharp lines of demarcation between mass units and their background of empty space. In the single entity concept, these apparent demarcations are just steep gradients of transition between different phases of the entity that are mathematically and conceptually misinterpreted as sharp boundaries.

Questioning the <u>qualitative</u> reality of concepts like electrons, protons, and electric charges should not be seen as a denial of, or failure to recognize, the valuable technologies that have obviously flowed from the <u>quantitative</u> values and relations that were derived using the standard atomic model as a guide. On the other hand, the advancements in such fields as electronic technology and nuclear energy physics should not be taken as automatic endorsements of qualitative reality concerning the intra atomic concepts that stimulated their development. So, conceptual flaws or not, the current atomic model has far too much practical and technological importance to be discarded. I would suggest, however, that we should constantly strive to develop more realistic alternative or parallel models, and then compare the ease with which all future observations and discoveries can be fitted into these different atomic concepts.

As a baseline for all atomic modeling, it has been generally (and properly) assumed that nature uses only one method of intra atomic organization that does not vary with time or location within the universe. This assumption is reinforced by the uniform quality of all stellar light that reaches us from different directions and distances. It also implies that atoms of hydrogen, carbon, or iron should be considered as identical to all other hydrogen, carbon, and iron atoms without regard to time or their particular galaxy of origin.

Clearly, there is a very impressive integral or whole number type of perceived unit division that is characteristic for the interior of each atom. Also, there seems to be uniformity in the physical features of atom building units that does not vary from one type of atom to another. Of course, the symmetry of the periodic table tends to confirm both of these last two statements.

The apparent discreteness of intra atomic building units has historically been pictured as a bi-phasic or two component system in which intra atomic mass units are sharply delineated and separated from each other as well as from their containing matrix of empty space. Consider, however, that it is possible for a single entity within the atomic volume to make use of variable gradients and organize in such a way that it is perceived as having uniform, discrete unit construction.

As mentioned before, I have no specific geometry in mind that would explain exactly how a single entity could organize to mimic our unitary perception of the atomic interior. So, the following discussion of single entity models is purely speculative. The purpose is not to picture the reality of the situation, but only to stimulate a search for the real geometry by pointing out a few organizational possibilities. We might begin by listing a few assumptive rules that can be followed in trying to visualize a more realistic atomic model.

(1) — There is no empty space inside the atom.
(2) — The volume inside the atom, like all volume, is completely occupied by various phases of the single universal entity.
(3) — This universal entity is in a natural and perpetual state of somewhat complicated curved motion that has gradient curvature in its several descriptive parameters.
(4) — The only force involved is a universal cohesion or resistance to separation. (The idea that two attracting units would collapse to one unless the attraction was counterbalanced by rotational centrifugal force does not apply in the single entity model, since the absence of intervening empty space obviates the concept of collapse.)
(5) — This cohesive force between two contiguous volumes of entity increases with average motion frequency and proximity of the volumes. So, the cohesive force is also inversely related to the average radius of curvature of the adjoining motions.

Nature obviously has a tendency toward hierarchy

organization on the macroscopic or cosmological scale, so it would be reasonable to expect a similar tendency at the subatomic level. Let's presume that the single universal entity tends to organize itself into several levels of hierarchy containment below the atomic volume.

For demonstration purposes only, we might consider the nuclear volume and the electron shell volume as being two contiguous levels in this hierarchy geometry. The recognized very strong cohesion of nuclear components (commonly called the strong nuclear force) would be a function of the motion parameters concerning frequency, radius of curvature, and proximity as previously noted. Then, going up the hierarchy scale to the electron shell level, the unit size would be much larger with a greatly reduced inter unit cohesion (commonly considered as an electrical type force at this level) resulting from a proportional change in the motion parameters of frequency, radius of curvature and proximity.

Of course, the reality of the intra atomic organization is probably quite different from the two level example used above. So, at the risk of over complicating the model, I will inject a few more blending speculations. It is probably more reasonable to presume that there are several telescoped or containment types of hierarchy levels within the atomic volume. In this case, the atomic number (or possibly the atomic weight) might roughly parallel, or be proportional to, the number of hierarchy levels. Also, there might be a gradient in the number of member units at each level. Lastly, the variable motion parameters that distinguish the members on two adjoining levels from each other would have a blending gradient at the junction between these two levels rather than a sharp line of demarcation.

From a slightly different perspective, consider that there is a wide spectrum in the value of each of the quantifiable aspects of structure and behavior within the atomic volume. For example, the wide range of unit size from the tiny sub nuclear realm to the atomic volume would be one of these quantifiable aspects. Another would be the wide ranging value of the cohesive force. Each of these quantifiable aspects might be graphically represented as a sloping line, however this line would not be straight. Typically, the shape would be wavy or characterized by oscillating curvature. Considering this wavy aspect of value gradients makes it easier to see how a single entity can give rise to our perception of discrete unit type of organization concerning both structure and behavior of the atom's interior.

From the preceding discussions and modeling, it is obvious that the concept of a single naturally occurring cohesive or attractive force is not compatible with the standard explanation in which there is a qualitative division of intra atomic forces into "electrical" types and "strong force" intra nuclear types.

Admittedly, this single cohesive force has several different perceivable phases and strengths, but they all come from the geometry of motion interrelationships that occur within the atomic volume. The several hierarchy levels of structural organization within the atom are accompanied by several levels of motion parameter values which in turn cause several levels of cohesive force values. Thus, the particular intra atomic level of origin accounts for the range of perceptions that motivates us to assign different names to the variants of the single cohesive force.

As a corollary to the above, the concept of electric charges deserves some criticism at this point. Probably, the entire intra atomic electrical system of charges should be replaced by the system in which there is only variable cohesion or resistance to separation, the strength of which is determined by motion parameters between contiguous units as previously described. There is no repulsive force, so the ideas of positive and negative charges, including attraction of opposites and repulsion of like charges, would all become obsolete.

As a purely hypothetical example of intra atomic geometric possibilities, we might consider the following; two contiguous rotating units rotate in opposite directions so that there is complementary meshing at the interface rather than interference. Then, a linear series of these paired units can circle into a toroidal structure. This would represent a kind of two dimensional hierarchy building. It is interesting that these toroidal structures would have to be composed of an even number of rotating units in order for all the unit interfaces to mesh rather than interfere. Also, four and six could be numbers of some geometric importance in atomic hierarchy organization, since these are the smallest numbers of units that can form an interface motion meshing type of toroidal donut. Of course, the next step in hierarchy building would be that two contiguous toroidal structures rotate in opposite directions so that there is meshing at their rotational interface. The relationship of these toroidal like circles could be contiguous as just described or it could be a type of containment in which smaller toroids form in the donut hole of larger level toroidal structures. Possibly, natural hierarchy organization employs some combination of both the contiguous and containment organizational principles. Of course, the reality of the intra atomic geometry would be vastly or infinitely more complicated than the toroidal hierarchies described considering natures three dimensions, and because all the volume within the atom must be involved in this motion. Recall that no empty space is conceptually allowed.

Following nothing more substantial than instinct in this matter, it would seem necessary to have frictionless meshing between all contiguous rotations in order for the idea of permanent motion as the natural state of the universal entity to be tenable. Consider that the range of possible directional differences at the interface between two motion units is zero to 180°. Zero would be complete meshing and 180 would be exactly opposite directions. Frictionless meshing would seem to imply that intra atomic organization should be close to, or no more than a few degrees away from zero in its inter-unit motion relations. At any rate, the range of these directions away from zero introduces a whole new system of geometric variables that could influence every aspect of motion and structural organization within the atom.

That is probably enough rambling on the elusive subject of an atomic alternative. The complicated and incomplete nature of the preceding discussions reinforces the earlier admission that I do not yet have a theoretical replacement for the standard atomic model.

Scientists seem to have a deeply ingrained habit of imposing "discreteness" or sharp lines of demarcation into their conceptual theories. This is reflected by the multiple names (electrons, protons, hadrons, quarks, etc.) that have been assigned to various volumes within the atom. Thus, the misleading idea that subatomic particles are naturally subdivided into geographically distinct units becomes more entrenched. However, this concept of particle subdivision runs into problems when the illogical nature of infinite divisibility is considered. This criticism is usually answered by stating that the smallest particle, which might be one of the quark family, is not divisible. Of course, this means that the quark must then be qualitatively different, in some unspecified way, from all other larger mass particles that are divisible. Thus, the divisible particle concept is somewhat lacking in logical tidiness.

It might be more appropriate to define "mass" as a certain threshold of human perception. In size or volume, this threshold could be somewhere in the vicinity of hydrogen atom size, so that all unit organizations of the universal entity above that size would be humanly perceived as particulate mass. Below that size level of human mass perception, the universal entity would be seen as a completely homogenous substance. So, perceived unit organization at the subatomic level would be based on variable motion patterns of this homogenous entity rather than particle division.

This concept of mass as a threshold of human perception might be more easily explained by using motion as a basis for the definition. First, consider that the rotational motion or frequency of entity units becomes less or progressively slower as we move up through size hierarchy levels from the sub nuclear toward the atomic volume. Somewhere, at about hydrogen atom volume, this rotational motion becomes static relative to the human observer. Thus, this matching of motions marks the threshold of particulate mass perception.

No attempt will be made to discuss the details of quantum theory in this writing, but there are a few general types of criticisms that come to mind. The idea of trying to sharply marginate any natural phenomenon into quanta or packets, whether it is packets of energy or units of the structural entity, runs contrary to the gradual gradient type of blending that is more characteristic of the observerable macroscopic structures and organizations of the universe.

As an analogy, consider a well formed hurricane in the Gulf of Mexico. As viewed from Mars, this storm would seem to be a very sharply marginated, evenly opaque disc or sphere. Earth observers, however, would recognize it as a complicated dynamic structure in the shape of a toroidal spiral. There is a gradually decreasing rotational gradient of air velocity from the center toward the periphery of the storm. At the periphery there is another gradient or gradual transition of air velocity down to the static condition of the air around the storm. At the center there is a third and steeper type of gradient that would be used to describe the transition of air velocity from the high inner core values down to the low eye of the hurricane values. So, discreteness at a distance may not hold up under closer examination.

Quantum theory and its sub headings of quantum mechanics and quantum cosmology is a field that accepts a great deal of vague, indefinite, and probabilistic types of information in building a shaky platform from which to launch its predictions. In Heisenberg's uncertainty principle, for example, a certain indefiniteness and uncertainty is considered as unavoidable. When it was recognized that position and momentum of a single particle could not be described simultaneously, the prevailing theoretical guide lines allowed this paradox to be carried forward using probabilistic and wave function explanations to rationalize the problem. I would suggest that it might be more logical to back off and re-evaluate the basic assumptions and definitions of momentum, position, and particle; in other words, re-think the structure, motion, and geometric organization of the intra atomic volume rather than justify and accept the contradictions of an obviously inadequate theory.

In the last 25 years or so the Black Hole scenario has become increasingly popular. In this theoretical model, objects more massive than several solar masses are destined to undergo sudden, almost unlimited gravitational collapse.

There is nothing wrong with the general concept that mass concentration can proceed further than the degree of compactness calculated for the largest stars. Concentrations of millions or billions of stellar masses obviously occurs at the core of many active galaxies (and quasars). This concentration is logically accompanied by an increase in temperature and gravitational compression as well as a decrease in atomic volume occupancy that are all magnitudes greater than the comparable values existing in stellar cores. I would suggest, however, that these value changes proceed in a more or less straight line fashion rather than passing through a precipitous threshold of collapse. The mathematical extrapolation that predicts sudden gravitational collapse is based on the faulty assumption that there is a great deal of empty space within the atomic volume. The mythical resistance to collapse is overwhelmed and everything implodes through empty space to a volume that is almost unlimited in its smallness. However, in the more probable atomic model involving a single entity with no empty space between contiguous units, the idea of collapse becomes meaningless; thus the popular and boldly unconventional black hole concept lacks a realistic mechanism of formation.

FORCES AS VARIANTS OF UNIVERSAL COHESION

Considering the universe as a diversified or many faceted event, then applying some organizational imagination, we can logically divide that event into three general categories of reality as follows:

> One category is the single universal entity or substance that exists in its very widely perceived range of mass and space forms.
>
> Second is the constant multi-directional electromagnetic wave disturbances (light for short) that propagate through the entity with a speed of c.
>
> The third category, part of which is commonly called gravity, is the universal force of cohesion that resists shape change or volume increase in any of the units or volumes of the universal entity, which is the designated first category.

So, these three categories that combine to make the universal event are extremely different. One is a substance, the second is a wave disturbance of that substance, and the third is neither a substance nor a wave disturbance, but an omnipresent static condition of variable affinity.

There is an understandable human tendency to simplify or consolidate these reality categories by designating constants or common behavior patterns to bridge the gap between the three classes of phenomena. One example of this inappropriate unification is the attempt to assign particle or quantum features of the substance category to the disturbance category. Another is the imposition of standard disturbance velocity values as a limit for inter substance velocities. Entirely too much category bridging importance has been attached to the value of c. More logically, as previously discussed, c should be regarded only as the standard (or small variable range) speed of spherical wave disturbances through the space phase of the entity. In addition to being theoretically linked to each of the three reality categories, c has been used as a constant to justify various combinations of the three.

In another example of questionable reality combination, general relativity contains the assumption that describes a mutual value shaping relationship between gravity and the theoretical space-time entity. It seems that relativists consider that this category mixing of gravity, space-time inter dependence is a minor transgression on our logical sensibilities (like adding oranges and apples) that simply must be accepted to satisfy the more compelling need for a beautiful, symmetrical unification among all aspects of nature.

The preceding philosophical comments notwithstanding, the reader can be reassured that this chapter will eventually turn to the designated discussion of forces and universal cohesion — after a few more introductory generalizations.

In contemplating natural phenomena there is a very helpful state of mind that might be described as being receptive to simple plain vanilla behavior that has absolutely no premeditation. The universe is clearly such a non thinking entity that is behaving like a runaway train following the path of least resistance with no guidance or pre planning involved. It should be recognized, however, that with time, even a simple unmotivated process involving a minimum of raw materials can become accidentally and randomly complicated in a fashion that is well described by the principle of entropy. Such is obviously the case with today's universe.

The key to cataloging and understanding the origin of this myriad of universe events is the recognition of the importance of qualitative similarities while not being stampeded by geometric and quantitative variables into separately naming events that should more realistically be grouped as various phases of a single type of behavior. Attempts to unravel the mystery of what causes this variety of motion events has resulted in separate names for the forces that power the various events. *(Multiple forces for various events is faintly reminiscent of the ancient habit of assigning separate deities to various weather conditions.)* It would seem logical to attach more importance to qualitative similarity found in the cohesive or attractive nature of the four named forces, rather than to allow differential naming to prevail because of the wide ranging quantitative parameters in such things as strength and distance of action.

In mathematical use, force is taken as a single quantity, with no consideration of the fact that the term implies either a pushing or pulling type of activity. This is a potential source of confusion, since the four named natural forces are all cohesive or pulling in type. Furthermore, any separation or expansive type of motion that seems to be caused by natural repulsion can most probably be explained as the result of differential cohesion. This is described more fully in the bar magnet example at the end of this chapter. Taking this principle a step further logically suggests that all movement in the universe in all directions, including separations, closures, explosions, orbits, etc., is either directly or indirectly the result of some aspect of universal cohesion.

An initial definition of gravity that should be acceptable to most would be the cohesion or affinity of the mass or substance of the universe for itself. In earlier chapters I made more frequent use of the term "universal attractive force" or

UAF. For ease of discussion in the following material, let's consider that terms like gravity, universal cohesion, universal attractive force, and universal affinity will all be used interchangeably under the general definition of gravity as stated above.

The origin of the four named attractive forces is a subject that has been previously modeled and argued in the chapter on subatomic structure. Unfortunately, the unobserveable nature of the subject necessarily reduces it to little more than a group of speculations guided by imagination and intuition which can be summarized in the following way. The named attractive forces should logically be considered as only perceptual variants of a single type of force having its origin in the motion relations of the several levels of hierarchy organization within the atom. Also, the range of cohesive strength values of this force parallels the intra atomic levels of structural organization.

For convenience and clarity of discussion and for the purpose of keeping long descriptive terms to a minimum, the intra atomic organizational hierarchy will arbitrarily be divided into four designated levels.

> Level 4 is the entire atomic volume.
>
> Level 1 is synonymous with the standard concept of a nucleus populated by protons and neutrons. The characteristics at this level are very close approximation of the composing units with unit motion that has the greatest frequency and the smallest radius of curvature. The great cohesive force at level 1 is a consequence of these two characteristics.
>
> Level 2 would correspond to the standard concept of electron shells and rotational orbits or energy levels with a range of cohesive strength that is commonly described

in terms of positive and negative electrical interactions. Compared with level 1, this level has much less inter unit cohesion that accompanies the vastly different motion parameters. This level, like all levels, might more realistically be pictured as a poorly demarcated series of sub levels whose motion and structural parameters blend into each other.

Level 3, for easy identification, will be reserved for the particular geometries of unit organization that cause magnetic attraction.

The arbitrary and unreal features of this four level distinction should be emphasized. Natural random organization most probably builds an atomic structure that has a greater and variable number of barely discernable levels of organization. The standard picture of atomic structure, however, strays even further from this natural principle of gradients and parameter blending by assigning very distinct values for mass, volume, and motion (both spin and rotational) to items like electrons, protons, neutrons, orbits, and electric charges.

By way of introducing a comparative discussion, gravity might be identified as the distant manifestation of the cumulative or additive effect of the level 1 phase of intra atomic cohesion. Similarly, naturally occurring ferromagnetism can be seen as an additive type distant manifestation of the symmetry and alignment in the pattern of unit organization at intra atomic level 3. *(For comparative purposes it is convenient to use ferromagnetism as a typical example of naturally occurring attractive or cohesive force within the larger category of magnetism.)*

It is generally accepted that the origin of ferromagnetism is the rotational activity of contiguous items within the atom.

Iron and a few other elements are unique in that their level 3 items or units can organize into stable geometric patterns that result in mutually inducing directional similarity plus a distant additive effect of their combined attractive forces. It is also understood that this orientation within one atom causes a similar orientation to occur within neighboring atoms. In effect, this represents influence or action at a distance, since these intra atomic patterns are not in direct contact with each other. Next, domains of thousands of atoms develop in which they all have their level 3 geometries similarly aligned. Then, the cumulative effect of multiple domains allows groups of domains to act as a unit across a much larger space to have an aligning influence as well as an attractive action on similar groupings in neighboring iron or permanent magnets. Thus we have described action at a distance in which this attractive force spontaneously organizes itself into at least two levels of hierarchy containment above and beyond its level 3 site of origin.

Variable reaction to temperature is one way to demonstrate that the two distant effects of gravity and ferromagnetism originate from two functionally independent levels within the atom. Heating a permanent magnet to its Curie temperature of about 1400°F disrupts the level 3 geometry within atoms of a magnetic domain so that ferromagnetic attraction is lost. Significantly, though, the weight or gravity potential of the magnet is unaffected, reflecting the lesser temperature susceptibility of level 1 organization. However, raising the temperature to several million degrees would similarly disrupt the motion geometry at level 1 by causing fusion reactions to begin. This results in a loss of some mass, hence a reduction in weight and a lessening of distant gravity effect exerted by the test object. Therefore, these two examples suggest an inverse relation between the threshold temperature of cohesive disruption and the designated numbers of intra atomic levels. In other words, the smaller the intra atomic level, the greater the temperature that is needed to disrupt unit geometries and motion related cohesive behavior.

In summary, gravity and ferromagnetism are just two different reflections of, or distant manifestations of, two different intra atomic hierarchy levels of motion related cohesion among unit members at those levels.

As a preliminary to discussing changes in the strength of cohesion, recall the previous assumption that the natural and permanent state of all the smallest organizational units of both the space and mass phases of the universal entity is one of high speed curved motion. Also, that the strength of cohesion at all locations in the universe is a function of the various parameters of this curved motion at those locations.

By arbitrarily eliminating explosive and expansive events from a thought model, we can say that the pattern of cohesive strength is static throughout the universe. Even though there is cohesive strength variation from one location to another, there is no change from one moment to the next. Another example of this situation would be two magnetic objects permanently held about one inch apart. The level of cohesion or attraction between the two is permanent and constant. There is no wave propagation of any kind, like energy or attractive force waves, as long as there is no change in the internal motion parameters responsible for the mutual attraction between the two objects.

Now, putting explosions and expansions back into the model, we can say that whenever there is a local change in the strength of cohesion from whatever cause, including natural explosions and expansions, the message of this change expands from the location as a spherical wave. The message is a propagating disturbance, possibly involving a shape change to the units of the space entity that are in contact with the changing cohesive event. Apparently, the message of a local strength change in gravity rides the same medium that light uses in its spherical wave propagation. There obviously must be a qualitative difference in the types of disturbances that are introduced into the space medium by these two different phenomena of light and cohesive strength change, but the transmission speed would logically be the same in both cases since they use the same medium. So, I would agree with the concept of gravity waves at the speed of light as long as it is limited to describing the propagation of changes in gravity rather than being applied to the situation of unchanging cohesion between two objects.

As a somewhat practical example of the above gravity discussion, consider that on earth in 1995 we became visually aware of a super nova explosion some 1000 light years distant. This explosion was characterized by considerable mass disappearance, thus the event included diminution in the local strength of cohesion or gravity. So, if we had instruments that could detect small decreases in cohesive pull from narrow directions, they would probably register these changes in 1995 simultaneously with the light message from the super nova, since both phenomena use the same intervening space phase entity to transmit their qualitatively different media disturbance messages.

There is a small scale experimental situation that seems to be an appropriate analogy to the wave transmission of gravity strength changes. First, consider a model in which the artificial high speed rotation of two permanent magnets (in the fashion of an electric generator) would cause a rapidly fluctuating change in the strength of cohesion concerning the effect on the immediately surrounding space entity units, which would be analogous to the gravity diminution effect of the super nova on its surrounding medium. So, the rotating magnet induced cohesive changes would wave propagate away from the event with a speed of c just like the gravity - super nova example. If there also happened to be a copper wire in the vicinity of these rotating magnets, then this changing cohesive strength would induce a disturbance or shape distortion in the motion geometries at one or some of the copper atom's intra atomic hierarchy levels. This disturbance would propagate at the speed of c from atom to atom along the length of the wire. Of course, this phenomenon is commonly called electricity resulting from magnetic induction.

The implication from this magnet analogy is that electricity in a wire and the wave propagation of cohesive strength changes away from the event are the same process. Perceptual differences arise in this matter because the structural organization of copper happens to form a very efficient directional geometric containment for the propagation of this disturbance.

Another probably valid generalization would be that the mechanism of propagation is basically the same for light, electricity, and cohesive strength changes; also, that the perceived difference between electricity and light is due to a qualitative difference in the nature of the disturbance that is imposed on a common type of transmission medium. Possibly such terms as "transverse" and "longitudinal" might be useful in trying to describe the directional aspects of the disturbances in motion geometry and shape involved in these two perceptual variants of wave propagation.

In the early part of this century from about 1905 to 1920 or so, Einstein, with minimal help from others, developed the general theory of relativity in which gravity was theoretically related in various ways to other phenomena such as acceleration, time, space-time, and light transmission. Having some criticism of general relativity assumptions and conclusions, I will try to explain my reservations without retracing this entire somewhat involved 15 year sequence of observations, logical arguments and deductions that comprise the building of the theory. There are, however, several frequently cited observational explanations and relativity event grouping situations that are integral to the logical development and credibility of general relativity. To keep this section reasonably short, I will confine my critical discussion to pointing out alternative ways to interpret these few events.

Probably the zenith of popular and scientific community acceptance of general relativity came with the 1919 solar eclipse experiment in which it was observed that starlight was slightly deflected or bent when it passed near the sun in route to earth observers. This observation was previously predicted as part of general relativity because of the assumption that gravity should have a direct attractive effect on light rays. There is, however, another way to explain bending of light waves around such a gravity strong object. Recall my previously explained concept of space as a phase of the universal entity that is affected by gravity in its motion status as well as by having a spectrum or gradient of density that is gravity strength related just like the more dense mass phase of the entity. Light would transmit more slowly in the denser space medium immediately around the sun resulting in a prism or refraction like effect that would give the observed small angle of light bending. A very close analogy (not an exact identity of event determinants, but an analogy) to the solar situation would be the way that earth's atmosphere uses its gravity induced gradient of light transmission speed to produce the color, size, and position variants that shape our perceptions of sunrises and sunsets. As a general observation, it clearly seems more reasonable to picture gravity as attracting a substance or entity like space phase entity rather than having it pull on an event or disturbance like light waves.

In some relativity accounts, the observation that the wave length of solar light is slightly red shifted seems to be erroneously attributed to the solar gravitational effect on the light that is in transit between the sun and earth. Without considering or accepting the general relativity assumption of a gravity, space-time interdependent relationship, the following hypothesis can be made. Wave length can only be lengthened by reducing the frequency at the point of origin on the surface of the sun. Logically, this would seem to be something that parallels the greater atomic gravitational compression effect on the sun. Recall that this type of red shift mechanism was also used earlier to explain the puzzling non-cosmological red shifts of gravitationally strong objects like quasars. So, light in transit from the sun to earth would only be temporarily slowed in accordance with the gravitationally induced density gradient of the space medium surrounding the sun. Earth observers, however, would detect the usual speed of light because light as a media dependent disturbance would revert to the speed consistent with local conditions of media density and media movement as it approached earth. Thus, on earth, we receive sunlight at the standard speed of c and with a tiny red shift that was inserted at the source before the trip began.

The problem of explaining why Mercury's elliptical orbit seems to rotate slightly faster than it should is something that leads very quickly to complicated math and speculations that I prefer to avoid at this time. As a general comment, I would only suggest that both the motion status and the density gradient of the solar system's background matrix of space phase entity are probably significant factors in explaining the precession of Mercury's perihelion as well as tiny orbital aberrations for all planets.

Observations in the 1960s (R.V. Pound and students) were interpreted as a gravitational red shift between the top and bottom of a tower at Harvard University. Supposedly, this red shift was caused by the gravitational effect on the light speed pulses traveling from the bottom toward the top of the tower. I would suggest as before, however, that decreased frequency from increased gravitational compression of the emitting device (probably an atomic clock) at the bottom of the tower would be a more realistic explanation for any detected red shift. Of course, there is a very tiny increase in the speed of light from the bottom to the top due to the decreasing density gradient of the transmitting medium within the tower. So, by adhering to the constancy of the speed of light (as in special relativity postulates) it is possible that investigators could misinterpret observational measurements reflecting light speed variations as wave length changes.

In summarizing the problems of general relativity, it should first be noted that there is no argument with the numerous examples of mechanical equivalence of gravity and acceleration. It is only when the desire for a completely holistic and beautifully symmetrical picture of nature motivates unrealistic dependency relations between independent phenomena that the theory runs into trouble. General relativity has flaws to the extent that special relativity assumptive errors are carried over. These include such things as faulty concepts of the nature, transmission, and speed of light as well as space medium — empty space misconceptions. There are misconceptions about the nature of gravity and its origin. The most obvious error is in the assumption that gravity can attract an event disturbance like light.

There is one peculiar and interesting aspect of naturally

occurring force that has received very little attention, possibly because it is not clearly understood. It is the repulsion between the like poles of two permanent magnets. This seems to be a singular exception to the attractive or cohesive nature of all other examples of naturally occurring force.

Recall at this point that there is directional orientation of the intra atomic units of the magnetic material, and that there is inter atomic influence of these level 3 units creating large domains of atoms with matching directional components; also, that both the North and South poles of a single magnet exert equal attractive forces on a piece of unmagnetized iron.

To set the stage for analyzing the repulsion between like poles we should first consider the effect of a bar magnet on a stationary piece of unmagnetized iron. The intra atomic magnetic units of this iron are randomly oriented more or less equally in all spherical directions. So, when the N pole of the bar magnet is pointed toward the piece of iron, all or most of the directionally coordinated (domain organized) magnetic units of the bar magnet are attracting only the fraction of the stationary iron units that happen to be oriented close enough to the proper direction for receiving that N pole type of attraction. Next, the bar magnet is reversed so that the S pole is directed at the unmagnetized iron. The overall attraction between the two bodies is the same in both N and S bar magnet positions, and the fraction of the stationary iron units involved is somewhat less than half in both situations, but the internal realities are quite different. With the S pole of the bar magnet facing the stationary iron, the participating units of the iron are not the same ones that were involved in the previous bar magnet position. The important point is that the N and S magnet positions dictate that two entirely different populations of atoms within the piece of iron are involved in these two attractive situations.

Now, consider the attraction when the N and S poles of two bar magnets are approximated. The attraction between the two objects is something in the range of twice the strength as in the previously described case. The reason for this is that the attraction now involves most of the units in both objects rather than less than half as in the case involving unmagnetized iron. Reversing the position of one magnet so that the two N poles are facing each other would logically be expected to reduce the attraction to near zero. This does occur, but the perception of the situation is confused because there is an apparent repulsion between the two poles that also develops.

Understanding this repulsion requires another brief digression as follows. An isolated object or bar magnet is apparently acted on by a surrounding sphere of universal cohesion. The object remains somewhat stationary because it is being pulled equally in all spherical directions by this cohesion or universal attractive force.

So, when the N poles of the two bar magnets are approximated, the expected drop in attraction occurs. Also, there is apparently a mutual shielding of the two magnets against a narrow segment of the surrounding universal cohesive force that normally operates along the linear direction of each bar magnet. As a clarification, picture a model in which the observer is facing two linear bar magnets that are aligned horizontally with the two N poles directed toward each other. The left hand magnet would then be shielded or somehow prevented from receiving the narrow cylinder of universal cohesive force that is in line with the linear axis of the magnets and coming from the right side. This means that the left magnet would tend to move toward the left in response to the unopposed universal attractive force from that direction. Of course, the reciprocal of this activity would cause the right magnet to move to the right.

I do not yet have any strong opinions about the mechanism of this mutual blocking of universal cohesion by the two magnets. Possibly there is some kind of shadowing or shielding by the right magnet that prevents the straight line projection of universal cohesion from passing through the body of the right magnet to reach the left one. Or there might be some kind of interaction between the right magnet and the right side cohesive force that usurps or reduces the amount of this universal cohesion that gets through the right magnet to the left one.

Of course, I have no evidence for this sudden revelation that these magnets, and all mass for that matter, are constantly being pulled equally in all spherical directions by a background of universal cohesion. I would only suggest that its existence furnishes a very logical and straight forward explanation of repulsion between magnets, in addition to being consistent with the previously developed concept of a single attractive force binding together all phases of a single universal entity.

An overview of naturally occurring force reveals that all aspects of the phenomenon seem to be attractive or cohesive in nature. Thus, the compelling weight of logic tells us that the repulsion between two N magnet poles must represent some perception of differential attraction rather than an isolated aberrant repulsive force.

Musing on the implications of this surrounding universal cohesion, it comes to mind that it might be experimentally possible to produce motion of mass by using this background of universal attraction in a differential or directional fashion. Of course, such a demonstration would probably come very close to the elusive perpetual motion machine. Possibly some reader will be motivated to imagine an appropriate experiment for this situation.

EPILOGUE

In closing this series of theory discussions and alternative models, there are a few final summarizing observations and philosophical comments that can be made.

The manuscript material might be summarized in a nutshell fashion by noting that most of the chapters originate from either one of two basic ideas that are contrary to generally accepted theory.

The first idea category involves the substitution of a very old and infinitely large static universe for the expanding model with its limited size and age. Then, considering an ancient universe, today's dynamic activity on all size levels from atomic fusion to galaxy organization suggests that there must be perpetual self sustaining cycles to explain the long term persistence of chemical and gross structural equilibrium. So, it is the logical flow from this idea category that gives rise to the topics and chapters on such things as equilibrium, hierarchies, size, age and motion of the universe.

The other basic idea in this nutshell summary concerns the denial of "empty space," which implies that some phase of a single universal entity should occupy the entire universe volume. This core assumption logically motivates the other chapters and topics like subatomic structure, universal entity, light transmission, special relativity, and forces as perceptual variants of universal entity cohesion.

In my opinion, both of the above renegade assumptions should be considered as very high in reality probability. Of course, this opinion is not yet generally accepted, but there is a growing number of scientists who would go along with the first idea category opposing the Big Bang sequence. Because of this, I feel that it is just a matter of time until observations and evidence topple the finite expanding model and thus allow other universe parameters to be considered. However, in the matter of empty space versus universal entity, even though the probability is strong for the single entity answer, the observations and evidence that could turn scientific and public opinion in that direction may be very elusive. It may or may not be possible to uncover the knowledge that would eventually settle this question.

The term "common sense" has always been, and still is, a reliable baseline for human thought concerning practical and survival types of activity. However, human history is packed with examples of incorrect theories on the structure and behavior of nature resulting from the naive presumption that certain quantities and qualitative identities are too obvious to be questioned. Some things have historically been accepted as true just because it would seem to be foolish or nonsensical to do otherwise. This reliance on common sense and instinct has probably been an important ingredient in the evolution and development of the human species to its present status in spite of the blame it must accept for leading us into theoretical pitfalls. A quick example of over reliance on common sense would be the numerous successive limitations on the size of the universe based solely, time after time, on the range of vision or detection that happened to prevail in that particular historical era. However, contrary to the above thinking trend, there have also been many modern examples of abandoning common sense in favor of abstract ideas that are more dramatic and exciting. Such things as time dilation, differential ageing, universal speed limit, Big Bang theory, and black holes would be examples of ideas whose acceptance might be partially related to this intellectual mind set.

It appears that the general public tendency to accept, retain, or reject various theories may closely parallel the mood of the times. Choices sometimes hinge on whether it is fashionable or modern to lean more toward common sense or whether avant-garde thinking prevails and thus favors a more sophisticated espousal of shocking, bizarre or unconventional concepts. In modern history, there seems to have been a number of pendulum swings in the dominance of these two basically different intellectual attitudes.

The black hole sequence of events will be given a final critical summary at this time. Recall that the wide public acceptance of black holes is probably a reflection of the contemporary trend that leans more toward the shocking and abstract than the common sense end of the spectrum.

Criticism of the black hole concept can be conveniently separated into three classes of arguments. First, in an overview criticism, this object of extreme gravitational compression is not consistent with the idea of a very old universe in equilibrium. It is something of a dead end event with no apparent way to re-enter the perpetual cycle of activity that maintains equilibrium in an infinitely old universe.

The second type of argument will only be summarized briefly since it was covered earlier in more detail. A single entity universe with no intervening empty space is not compatible with the black hole story because the sequence requires a great deal of empty space within the atom so that precipitous and almost unlimited gravitational collapse can be a possibility.

The third criticism involves the frequently repeated and inappropriate analogy in which the intense gravity of the super dense black hole object prevents the "escape" of light.

As a background, in order to understand what is wrong with the above statement, consider the previously described "no empty space" model in which the interior of an active galaxy (or possibly a quasar) would have gradients of increasing temperature, increasing gravitational compression, and decreasing volume occupancy of all atomic and subatomic units as the situation is described from the periphery toward the center of the object. Of course, there would be hills, valleys and oscillations in the graphic representation of these density related gradients, but no precipitous drop offs. There would also be thresholds and plateaus concerning various classes of subatomic activity of which fusion is just one known example. Logically, there might be a qualitative difference in the type of micro activity occurring at the super high density core of an active galaxy as compared to the fusion process known to be happening near the periphery. Hence, one would not necessarily expect the core activity to emit a spectrum of radiation resembling the type that is generated in peripheral fusion activity.

Getting back to the third black hole criticism, the fact that we receive little or no radiation message from the interior of compressed objects like galaxy cores or quasars might be explained as follows. The particular qualitative variant of micro activity occurring in such a dense core may or may not cause the usual sequence of a disturbance that is propagated by the entity units surrounding the core. This disturbance, if it occurs, may be reaching us in some form that is completely undetectable or possibly it is being blocked or absorbed by the shells of galactic mass that surround the core. At any rate, the above explanation seems to be more comfortably logical than the standard black hole mechanism of preventing the escape of light by having gravity pull on, or retard, the progress of an event or wave propagating disturbance such as radiant energy.

As it happens, the black hole and the Big Bang concepts have probably both enjoyed a great deal more general recognition and acceptance than they deserve, due in part to media attention, and partly because of the daring and flamboyant nature of their mechanisms and implications. Undoubtedly, terminology is also important in that the superficial appeal of dynamic names seems to have considerable influence on the popularity of ideas like black holes and the Big Bang. The catchy and alliterative nature of their titles almost certainly contributes to their frequent appearances in newspapers and magazines as well as their longevity in the memory of the general public. Concepts like these seem to ride a wave crest of general popularity fueled more by repetition operating as a badge of implied knowledge than the reality probability of their contents.

In recent decades, it seems that mathematics is becoming progressively more important in the process of theory building. Admittedly, trying to unravel nature's secrets generates a number of intellectual disciplines that need some degree of systematic formalism for their development. Just minimizing the incursion of undesirable qualities like emotion shaping, purposeful manipulation, and the herd instinct requires some adherence to a generally accepted scientific method. However, pursuit of this desirable goal sometimes promotes an over reliance on mathematics.

Mathematical representation usually follows very quickly whenever a new theory or concept is put forward. There seems to be some sort of scientific investigative sequence or procedural ritual requiring that ideas must be translated into the language of equations before they can be stamped as credible. Then, if the equation can be molded into a symmetrical form or "solved," this accomplishment may be taken as evidence that the motivating concept must have been correct. Even though math is a valuable and necessary scientific tool, its basic limitation of being only a man made shorthand type of language should not be forgotten. Clearly, math is frequently misused in arguments of theory confirmation. Such things as outcome oriented constants and assumptions that go into some equations can greatly influence the predictions and mathematical verifications that emerge.

Mother mathematics is a powerful intellectual intimidator. When one of her disciples fills a blackboard with complicated equations containing a number of carefully crafted result meshing constants and assumptions, most people, including most scientists, will be awe struck into theory expounding paralysis by their inability to follow this higher order of thinking. Unfortunately, it is automatically assumed by most, that experts with this skill must also be more correct in all of their conceptual or non-mathematical thinking than the rest of us.

The human mind, having evolved in the somewhat spatially and temporally constrained environment of this planet, seems to naturally resist thinking about alien open ended concepts like infinity in any of its contexts, whether it involves size, age, or an eternal cyclic equilibrium that never runs down.

Uncomfortable as it may be, contemplating infinite time in either the past or the future is actually more reasonable than trying to explain a beginning or an end. If the idea of a creative deity is omitted from the discussion, the temporal beginning of the universe would seem to require the suspension of causality. Trying to retain causality involves a sequence in which the circumstances that caused the first appearance of the universe must be explained, followed by an explanation of its precipitating circumstances, which naturally leads to an impossible and never ending series of explanations. The root of the thinking problem in this area might be better described as the tendency to assign "limits" to all aspects of reality like mass, time, speed, space, etc. There is nothing unreasonable about the human habit of cataloging the sequence of events by using past, present and future time designations as long as unexplainable limits like beginnings and ends are not allowed to gain an assumptive foothold.

Personally, as time passes, I am becoming more and more at ease with the idea of an infinitely old and infinitely large universe, as well as one that contains no empty space but only a single entity or substance whose total past and future potential behavior is the result of a single universal force of cohesion that has a very wide range of strength values and perceptions.

Without invoking some form of the anthropic principle or predetermination by an intelligent deity, the universe must be seen as an unexplained spontaneously existing entity. The entire activity of the universe must therefore be considered as randomly unconscious — just a perpetual passive response to some kind of natural and eternal background influence or force. Logically, the smallest building block subdivision of this universe complex should be a single substance. Anything more than one primary entity or substance would automatically suggest some kind of outcome oriented or teleological determinism.

Multiple entities and multiple forces are almost certain to be natural human modeling mistakes in that they were fabricated to explain the variety of structural appearances and behavior patterns that confront us. This has been done simply because it is not yet understood or appreciated that a complicated multi-tiered system of structural and behavioral diversity can be achieved by a single entity with a single type of activity if you give it unlimited billions of years to repeatedly cycle, mix, and use entropy to reach an equilibrium that is today's picture of the universe.

Concerning forces, the most obvious and important piece of unifying information is that the separately named forces are all attractive or cohesive in nature. Most theorists, however, prefer to elevate the quantitative and directional differences of forces in various settings to the more important position of being the identifying criteria. So, the standard view designates separate forces as the accepted norm and places the burden of proof on those who would prefer a single force. Possibly a reversal of this proof responsibility should be considered.